The theory of pattern formation, assumed to be applicable to all multicellular organisms, has been developed largely through the study of animal, and to a lesser extent, plant systems. Fungi, members of the third major Kingdom of eukaryotes, have not featured in these studies, although much research of fungal morphology has been undertaken with taxonomic intentions. This first account of the developmental biology of fungal morphogenesis considers whether evidence exists for the action of pattern forming mechanisms in the development of fungal structures. Chapters on the fruit body, on a range of aspects of the hyphae and the mycelium, and on genetic control and nuclear events in morphogenesis provide new insights into the mechanisms used in fungal construction. Aimed at developmental biologists as well as mycologists, the terminology has been kept as simple as possible to make the volume accessible to the widest audience.

PATTERNS IN FUNGAL DEVELOPMENT

PATTERNS IN FUNGAL DEVELOPMENT

Edited by

SIU-WAI CHIU and DAVID MOORE

CAMBRIDGE UNIVERSITY PRESS
Cambridge, New York, Melbourne, Madrid, Cape Town, Singapore,
São Paulo, Delhi, Dubai, Tokyo, Mexico City

Cambridge University Press
The Edinburgh Building, Cambridge CB2 8RU, UK

Published in the United States of America by Cambridge University Press, New York

www.cambridge.org
Information on this title: www.cambridge.org/9780521560474

First published 1996

A catalogue record for this publication is available from the British Library

Library of Congress Cataloguing in Publication data
Patterns in fungal development / edited by Siu-Wai Chiu and David Moore.
    p.   cm.
Based on papers from a symposium session held at the Fifth International Mycological
Congress, Vancouver, Aug. 1994.
    Includes index.
    ISBN 0 521 56047 0 (hbk.)
    1. Fungi – Morphogenesis – Congresses.   2. Fungi – Development – Congresses.
I. Chiu, Siu-Wai, 1961–   II. Moore, David, 1942–   III. International Mycological Congress
(5th: 1994: Vancouver, B.C.)
QK601.P37   1996
589.2′044–dc20   96–2343   CIP

ISBN 978-0-521-56047-4 Hardback

Contents

Contributors

Adrian N. Bourne
School of Biological Sciences, 1.800 Stopford Building, The University of Manchester, Manchester M13 9PT, United Kingdom

Gode B. Calleja
Diliman Institute, P-9 Dalan Roces, Area 14, UP Campus, Diliman, Lunsod Quezon, The Philippines

Siu-Wai Chiu
Department of Biology, The Chinese University of Hong Kong, Shatin, New Territories, Hong Kong

Byron F. Johnson
Department of Biology, Carleton University, Ottawa, Ontario K1S 5B6, Canada

Keith Klein
Department of Biological Sciences, Mankato State University, Mankato, MN 56002-8400, U.S.A.

David Moore
School of Biological Sciences, 1.800 Stopford Building, The University of Manchester, Manchester M13 9PT, United Kingdom

LilyAnn Novak Frazer
School of Biological Sciences, 1.800 Stopford Building, The University of Manchester, Manchester M13 9PT, United Kingdom

P. John Vierula
Department of Biology, Carleton University, Ottawa, Ontario K1S 5B6, Canada

Roy Watling
Royal Botanic Garden, Inverleith Row, Edinburgh EH3 5LR, United Kingdom

Bong Yul Yoo
Department of Biology, University of New Brunswick, Fredericton New Brunswick, Canada

Preface

Pattern formation characterises creation of the 'body plan' or, more formally, the distribution of differentiated cells or tissues in developing organs or organisms. Pattern formation depends on *positional information*, which instructs competent cells to differentiate in ways characteristic of their position in the structure. Positional information not only is provided by the environmental trigger but also is usually understood to be imparted by the concentration of one or more morphogens emitted from spatially distinct organisers. The basic rules of pattern formation seem to be that regional specification (directed by organisers producing morphogens) occurs first, regulating gene activity in ways specifically geared to morphogenesis so that particular cells are first specified (a state which is still flexible) and then determined (a state which is inflexible) to their differentiated fates. These statements seem applicable, in theory, to all multicellular organisms but they are derived largely from research on animal (and, to a lesser extent, plant) morphogenesis. The challenge taken up in this book is to establish whether evidence exists for such mechanisms in the development of fungal structures.

The great bulk of the published research on fungal morphogenesis has been done with taxonomic intentions. It has great value for its descriptive and comparative content, but precise developmental accounts are extremely rare and experimental approaches rarer still. Following an introductory description of the *prima facie* case for similar events (and perhaps similar mechanisms) to those found in other eukaryotes being involved in establishing the patterns inside a fungal fruit body, the individual chapters of this book deal with the hyphal growth mechanism; patterning in the mycelium; the genetic control of hyphal morphogenesis; nuclear events in morphogenesis; experimental approaches to the study of pattern formation; the case for hormone/growth factor involvement in hyphal morphogenesis; as well as phylogenetic and ecological aspects of tissue distribution patterns in fungal fruit bodies.

This volume is based on a Symposium session at the Fifth International Mycological Congress held in Vancouver in August 1994. The final stages in editing the text were largely completed during a visit made by DM to the Chinese University of Hong Kong. Thanks are due to the Leverhulme Trust for the award of a Research Grant which enabled this, the Department of Biology at CUHK for their hospitality, and The University of Manchester for leave of absence.

Siu-Wai Chiu
Department of Biology
The Chinese University of Hong Kong

David Moore
School of Biological Sciences
The University of Manchester

Chapter 1

Inside the developing mushroom – cells, tissues and tissue patterns

DAVID MOORE

Summary

Major tissues of the mushroom (cap, stem, veil, basal bulb, etc.) are established very early in development. This Chapter offers a generalised overview of mushroom morphology, morphogenesis and mechanics, with a view to identifying the controlling events which determine tissue differentiation and distribution. Currently we face a major problem of insufficient information on the detailed anatomy of the fruit body. Knowledge of the seat of growth which drives morphogenesis and appreciation of the balance between physical and biological phenomena are also often lacking. Application of simple numerical methods to microscopic analyses of the stem in *Coprinus* has produced a quantitative account of the dynamics of hyphal changes which sculpture even this, apparently simple, structure. In tissues of the cap, there is *prima facie* evidence for control of patterning by diffusion of chemical morphogens in morphogenetic fields. The generation of the agaric gill can be understood in terms of organising centres which interact with one another by production of (and response to) freely diffusing activator and inhibitor molecules. Recent analyses show that (i) agaric gills grow at their base, not their margins and (ii) agaric gills are initially convoluted, being stressed into their regular radial arrangement later.

Pattern developments in theory

The shape, form and structure of an organism arise through its morphogenetic development, but the basic 'body plan' of the developing structure (whether fungal, plant or animal) does not appear all at once. Instead, its shape and form are assembled through a series of distinct acts of

1

differentiation by cells already specified by earlier differentiations to belong to a particular morphogenetic pathway.

Cells differentiate in response to chemical signals from other regions of the developing structure and/or under the influence of suitable environmental triggers. These chemicals (none yet identified) may be termed organisers, inducers, morphogens, growth factors or hormones and seem to inhibit or stimulate entry to particular states of differentiation (Chapter 7). For example, morphogens may contribute to a morphogenetic field around a structure (cell or organ) which permits continued development of that structure but inhibits formation of another structure of the same type within the field. This can produce a pattern in the distribution of the structure over the tissue as a whole. If the inhibition caused by the morphogen is strong and extends (through its diffusion or active transduction) over a wide area, then the structures will be widely spaced. On the other hand, if the inhibitory influence of the morphogen is localised (perhaps because it decays rapidly) then the structures will be formed close together.

The morphogen is providing the cells which respond to it with positional information, which allows those cells to differentiate in a way characteristic of their position in the morphogenetic field. Positional information seems to be imparted by concentration gradients of one or more morphogens emitted from one or more spatially distinct organisers. Effectively the cell moderates its behaviour on the basis of the nature and strength of the incoming morphogen signals, consequently adjusting its response in accord with its position relative to the controlling organisers.

Phenomena such as these generate the spatial pattern of cell and tissue distributions which characterises the 'body plan' of the developing organism. They have been described most effectively in various aspects of animal and plant morphogenesis but evidence does exist that such mechanisms contribute to the development of fungal structures. Development, differentiation and morphogenesis involve change with time, and the dynamics of cell and tissue interactions in time contribute to the form and structure of the organism which results.

Morphogenesis has a time dimension

As fungal fruit bodies develop a complete change occurs in the behaviour of hyphae and hyphal branches. The invasive growth which characterises the vegetative mycelium (Chapters 2, 3), and which enables it to act 'as a mould' is drastically modified. Instead of diverging and avoiding contact with their fellows, hyphae of what will become a fruit body grow towards one another contributing, with their branches, to a population of co-operating hyphal systems which makes up the undifferentiated tissue described as the fruit body initial.

Regions of recognisable cap and stem can be resolved in microscope sections of even very small (mm size range or below) fruit body initials. Such images provide *prima facie* evidence for the pattern forming processes outlined above because formation of histologically distinct regions of these sorts requires that some organisation is imposed upon the homogeneous undifferentiated tissues. Over many years, Reijnders (1948, 1963, 1979) stressed the importance of: (i) the manner of development and nature of the veil and the 'epidermis' of the cap (called the pileipellis) in relation to protection of the developing cell layer which is responsible for eventually producing the (basidio)spores, (the hymenophore which carries the hymenium or spore-producing tissue); (ii) the sequence of development of the stem, cap and hymenophore, which are the major functional zones of the mushroom fruit body; (iii) the mode of development of the hymenophore. The terminology has been discussed by Watling (1985) and the taxonomic, phylogenetic and ontogenetic significance of these features are discussed by Watling & Moore (1994) and in Chapter 8.

Evidently, the major tissue regions of a mushroom fruit body are demarcated at a very early stage in development. For example, in *Coprinus cinereus*, cap and stem are clearly evident in fruit body initials only 800 μm tall (Moore, Elhiti & Butler, 1979) though this size represents only 1% of the size of a mature fruit body.

The most highly differentiated cells are found at the boundaries of tissue regions in such young fruit body initials (Williams, 1986). In the youngest specimens these perimeters are occupied by layers of parallel hyphae (called meristemoids by Reijnders, 1977) with short cell compartments, suggesting that these hyphae are involved in rapid septum formation (Chapter 2). But these are not meristems like those which occur in plants. Meristems do not occur in fungi. Nor do meristemoids imply that the tissue borders are actively growing outwards. Of course, the meristemoids are growing actively; at this stage in development, all regions are growing actively. But the most significant point is that the driving force for growth of the developing fruit body primordium is its internal expansion; growth at the boundaries of tissue layers takes place to compensate for increase in the area of the boundary layer which occurs because internal growth increases the volume of each tissue region.

Using *C. cinereus* as an example again, as a typical fruit body grows from

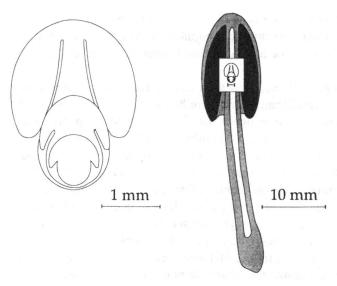

1 mm

10 mm

Fig. 1. Scale diagrams showing the size relationships of primordia and fruit bodies of *Coprinus cinereus*. On the left a set of three 'nested' outlines of vertical sections of primordia illustrate the steady outward expansion of the tissue layers. On the right, these same diagrams are superimposed (to scale) onto a median diagrammatic section of a mature basidiome to demonstrate the full extent of the outward movement of tissue boundaries. (Redrawn after Moore, 1995).

1 to 34 mm in height (i.e. a vertical linear change of 34 ×), the circumference of the stem in section increases 9 ×, the outer circumference of the cap in section increases 15 ×, but the volume increases more than 3000 × (Fig. 1).

   The differential growth which generates primordium enlargement exerts enormous mechanical effects on relationships between tissue layers which are often concentrically arranged. Mechanical forces themselves generate many of the patterns which characterise the form and structure of the mature mushroom fruit body. Some of these will be described in the following sections.

Co-ordination of cellular activities across the fruit body

The changes in shape and form which characterise fruit body maturation in basidiomycetes is usually described as 'expansion' because, clearly, the fruit body increases greatly in volume as it matures. Two basic strategies seem to be used to achieve fruit body expansion. The polypores and their close relatives (including those, like *Lentinula*, which have gills rather than pores) tend to produce more hyphal branches continuously; so they show no significant overall increase in cell size, but large increase in cell number. On the other hand the agarics depend on cell inflation for fruit body expansion and it is usually possible to show that the extent (and often the rate) of fruit body expansion far outstrips the production of new hyphae and/or branches.

There are two types of cell inflation. A slow process is usually found in young primordial stages, with a more rapid one being involved especially in stages of cap maturation. These sorts of cell inflation are distinct aspects of cell differentiation, representing functional specialisation of the cells concerned (Reijnders & Moore, 1985). Reijnders (1963) showed that the different parts of fruit bodies enlarge in proportion. The different tissues of the fruit body consequently develop and mature without being impeded or distorted by the growth of other parts. If fruit body primordia do enlarge proportionally the implication is that there may be some sort of co-ordinating mechanism operating over distances of many millimetres. On the other hand, an alternative interpretation might be that 'co-ordination' is more apparent than real because events may be arranged in a consequential series - one (secondary) event being instigated by the initiation or completion of an earlier (primary) event.

Cell inflation in agarics (as represented by various aspects of 'fruit body growth') has been the subject of a good deal of research, and reference should be made to Gooday (1974, 1982, 1985), Moore *et al.* (1979) and Rosin, Horner & Moore (1985) for work with *Coprinus cinereus*; Bret (1977) on *C. congregatus*; Hafner & Thielke (1970) on *C. radiatus*; Wong & Gruen (1977), Gruen (1982, 1991), and Williams, Beckett & Read (1985) on *Flammulina velutipes*; Bonner, Kane & Levey (1956) and Craig, Gull & Wood (1977) on *Agaricus bisporus*.

All of these studies examined individual processes in the fruit bodies concerned. There has only been one holistic account of inflation over the whole fruit body which enables an assessment of the correlation between cell behaviour in widely separated parts of the same fruit body (Hammad *et al.*, 1993a). The aim of this study was to determine how different tissues of the fruit body of *Coprinus cinereus* expanded both along the developmental time scale and in relation to the other tissues of the fruit body. Any such effort to examine co-ordination of events requires an objective time base. Many earlier studies have relied (quite satisfactorily within the context of the experiments undertaken) on morphological aspects of fruit body development to define the developmental stage of each specimen (Madelin, 1956; Takemaru & Kamada, 1972; Matthews & Niederpruem, 1973;

Table 1. *Designation of developmental stages of fruit body development in* Coprinus cinereus

Stage	Time	Predominant feature (reference a)	Other designations in the literature			
			b	c	d	e
1	0–12h	Basidial differentiation	Stages 1–2	Stages 1–3	Day 8	
2	13–24h	Karyogamy and meiotic prophase	Stage 2	Stage 2		−2 Days
3	25–36h	Meiotic divisions and spore formation	Stages 3–4	Stage 4	Day 9	−1 Day
4	37–48h	Stem elongation and spore release	Stage 5			Day 0

References: (a) Pukkila *et al.*, 1984; (b) Moore, Elhiti & Butler, 1979; (c) Morimoto & Oda, 1973, 1974; (d) Kamada, Kurita & Takemaru, 1978; (e) McLaughlin, 1982.

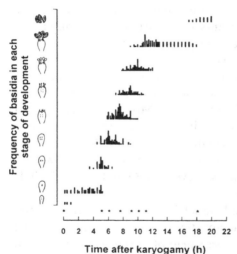

Fig. 2. Time course of progress through meiosis and sporulation in *C. cinereus*. The abscissa shows elapsed time. Samples were removed from fruit bodies at regular time intervals and sample populations of hymenial cells observed in squashes were quantitatively categorised into the developmental stages represented in the cartoons on the ordinate. Time 0h is arbitrarily set to the start of karyogamy. Other timings emerged from the observations (as the medians of the distributions shown): meiosis I occurred at 5 h, meiosis II at 6 h, sterigmata appeared at 7.5 h, basidiospores started to appear at 9 h, began pigmentation at 11 h and mature basidiospores were being discharged from 18 h. The asterisks on the abscissa emphasize these stages. (Redrawn after Hammad *et al.*, 1993a).

Morimoto & Oda, 1973; Moore *et al.*, 1979). Pukkila, Yashar & Binninger (1984) introduced nuclear staining to define the stage of development of the whole fruit body in terms of progress through meiosis and spore formation in the hymenium (Table 1). Hammad *et al.* (1993a) extended this approach by examining a sufficiently large sample of fruit bodies to establish the exact timing of major meiotic and sporulation events, creating a time base to which any other process can be referenced simply by microscopic examination of a sliver of cap tissue (Fig. 2). The resultant time base is objective because it depends on observation of processes which are endogenously controlled and reliable because the processes concerned are central to the functioning of the fruit body. Time courses could be established with individual fruit bodies because such small amounts of tissue were required for the staining technique used (Pukkila & Lu, 1985) that repeated samples could be removed at known time intervals.

Hammad *et al.* (1993a) performed a morphometric analysis by measuring cell sizes in microscope sections of fruit bodies whose stage of development was defined by the stage they had reached in meiosis and sporulation. Measurements were made of cells in different positions in the stem (base, middle and apex) and of a range of hymenial cell types (basidia, spores, paraphyses, cystidia and cystesia) and veil cells. The resultant sets of measurements were then analyzed for information about stem elongation and cap expansion; stem structure in longitudinal sections and the co-ordination of cell expansion.

Stem elongation and cap expansion

In the early phase of slow stem growth the rate was in the region of 10 μm min^{-1}; in the rapid elongation phase the growth rate was 110 μm min^{-1}. During the most rapid phase of elongation, an average stem may elongate 80 mm in less than 12 h (Moore & Ewaze, 1976). Rapid stem elongation in *C. cinereus* occurred in a phase which occupied most of the 5 h prior to spore discharge, starting 8 h after karyogamy. Cap expansion started as spores matured, about 14 hours after karyogamy (Fig. 3). Stem elongation was significantly greater when the cap was left attached than when it was removed (Table 2) and this was true whether the caps were removed before (10 h after karyogamy) or after (12 h after karyogamy) onset of the phase of rapid elongation. Leaving a half cap in place was sufficient to ensure normal elongation. These stems curved away from the side with the half-cap on it during the 8 h immediately following treatment (height range 80-85 mm),

Table 2. *Effect of the cap on stem elongation in* Coprinus cinereus

Treatment	Initial length (mm)	Final length (mm)	Elongation (mm)
(a) caps removed 10 h after karyogamy (just before onset of rapid elongation phase)	35.8 ± 4.9	109.2 ± 12.2	73.3 ± 8.8
(a) control	36.0 ± 5.5	130.4 ± 2.99	94.4 ± 6.3
(b) caps removed 12 h after karyogamy (just after onset of rapid elongation phase)	35.5 ± 5.11	23.0 ± 22.9	87.5 ± 19.3
(b) control	37.6 ± 5.4	155.6 ± 16.0	118.0 ± 13.7
(c) half of cap removed 12 h after karyogamy	47.0 ± 9.4	121.5 ± 14.2	74.5 ± 9.2
(c) control	48.0 ± 10.8	124.0 ± 12.8	76.3 ± 3.9

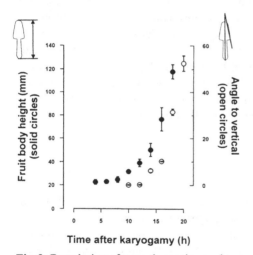

Fig. 3. Correlation of stem elongation and cap expansion with meiosis in *C. cinereus*. (Redrawn after Hammad *et al.*, 1993a).

but in the next 4 h, during which they grew to 104-137 mm in height, growth was strictly vertical.

The kinetics of stem elongation has attracted a great deal of attention. Buller (1924) examined *C. sterquilinus*; *C. radiatus* has been studied by Borriss (1934, using the name *C. lagopus*), Hafner & Thielke (1970) and Eilers (1974); *C. congregatus* by Bret (1977); and *C. cinereus* by Gooday (1974), Cox & Niederpruem (1975, using the name *C. lagopus*) and Kamada

Table 3. *Sizes of stem cells (in terms of sectional area) measured in longitudinal sections of stems and the length/width ratios of the cells*

Fruit body height (mm)	Section	Mean stem cell area (μm^2)	Length/width ratio
3	Middle	148	1.9
8	Middle	211	2.0
25	Apex	292	1.8
	Middle	3857	11.5
	Base	2705	6.2
48	Apex	3184	9.0
	Upper middle	6813	12.7
	Lower middle	5735	10.3
	Base	3449	10.6
55	Apex	9243	11.5
	Upper middle	9496	18.1
	Lower middle	10522	19.9
	Base	6533	13.8
83	Upper apex	6258	12.6
	Lower apex	11960	22.0
	Middle region 4	12894	26.0
	Middle region 3	11672	30.2
	Middle region 2	10448	35.1
	Middle region 1	5538	20.4
	Upper base	4785	14.1
	Lower base	2681	8.8

Each entry represents the mean of 50 measurements

& Takemaru (1977, using the name *C. macrorhizus*). In all these species the upper half, and generally the upper mid-region, has been shown to be the most active zone of elongation, and this part of the stem showed the most dramatic increases in cell size (Table 3). The species represented in the literature differ in the degree of stem autonomy and the relative parts played by cell division and cell elongation in the process of stem elongation. In *C. congregatus* stem elongation was dependent on both the parental mycelium and the cap during the whole period of development of the fruit body (Bret, 1977). In *C. radiatus* elongation occurred after separation from the parent mycelium, but was dependent on the cap only until the final phase of rapid elongation was reached: after the stem had reached 25% of its final size decapitation did not impair elongation (Borriss, 1934; Eilers, 1974). In contrast, Gooday (1974) showed that stem elongation in *C. cinereus* had no requirement for connection either with the cap or the parental mycelium. This was confirmed by Cox & Niederpruem (1975) who extended the observation by showing that primordia about 5 mm in height (which is equivalent to between 5% and 10% of final size) were able to elongate after excision and decapitation. Hammad *et al.* (1993a) demonstrated another aspect of cap/stem interplay by showing that the *C. cinereus* stem does not depend on the presence of the cap, but stem elongation certainly benefits considerably from the presence of the cap (Table 2).

Stem structure in longitudinal sections

The data in Table 3 show that there was little increase in the areas of cells in longitudinal sections between a 3 mm fruit body and an 8 mm tall fruit body, both of which were at pre-meiotic developmental stages. Since the mean cell length does not change while the fruit body more than doubles in length, it is presumed that stem elongation at these stages is due primarily to cell division rather than cell elongation. It is difficult to see how hyphal apical growth could contribute to this throughout the stem, and remembering that here we are discussing a structure made up of fungal hyphae, it might be speculated that 'cell division' in this context could take place by intercalary septum formation in the hyphae, followed by elongation of the daughter compartments to the size of their mother compartment. In contrast, there was a large increase in longitudinal sectional area of stem cells between the stems of the 8 mm fruit body (pre-meiotic) and that of a 25 mm fruit body (which was undergoing meiosis). Initially it was the cells in the basal and middle regions of the stem which inflated. The apical cells did not expand to the same extent. Even in a fully elongated fruit body apical cells were considerably shorter than cells in the other regions of the stem. The most elongated cells were found in the upper mid-region of the stem.

Cell length/width ratios did not change much in pre-meiotic stems (3 mm and 8 mm fruit bodies) but, particularly in the upper middle regions of the fruit bodies, increased greatly after meiosis. In the 48 mm fruit body stem cells were approximately 10 times longer than they were wide, in the 55 mm fruit body the cells in the upper middle region had a length/width ratio of approx. 20 and in the 83 mm tall fruit body the cells in the middle region of the stem this ratio approached 35 (compare with a ratio of 2 for cells in the same regions of 3 mm and 8 mm fruit bodies).

Stem elongation in *C. radiatus* was accompanied by a 6- to 8-fold increase in cell length and a doubling of the cell number (Eilers, 1974). In contrast, in *C. cinereus*, although the DNA content of the stem has been found to increase abruptly just before the most rapid phase of elongation (Kamada, Miyazaki & Takemaru, 1976) and stem cells become multinucleate (Lu, 1974; Moore *et al.*, 1979; Stephenson & Gooday, 1984), stem elongation has been attributed solely to cell elongation (Gooday, 1975; Kamada & Takemaru 1977). The data of Hammad *et al.* (1993a; see Table 3) support this view. Only between the 3 and 8 mm tall primordia was the increase in overall size greater than the increase in mean cell length. The increase in size of stem throughout the rest of the size range was easily accounted for by the increase in cell size.

Co-ordination of cell expansion

Hammad *et al.* (1993a) measured the sectional areas of cells of the hymenium in the same longitudinally sectioned specimens on which the stem cell measurements were made (Table 4). Large scale expansion of stem cells commenced at or just after meiosis (25 mm fruit body), this being reflected in the observation on the macro-scale that rapid stem elongation was correlated with the ending of meiosis (Fig. 3). Remarkably, expansion of the different cell types in the fruit body cap as well as inflation of cells of the stem began immediately post-meiotically (Table 4). Expansion of the cells of the stem is necessary to elevate the cap into the air for more effective spore dispersal; expansion of the different cell types in the gill tissue is also necessary for effective spore dispersal and co-ordination with stem expansion is clearly advantageous.

Hammad *et al.* (1993a) clearly demonstrated that the major phase of cell inflation was a post-meiotic event throughout the fruit body and could account for all the growth involved in fruit body maturation of *C. cinereus*. Also, inflation of cells in the cap was closely correlated with inflation of cells in the stem.

It is possible that the different tissue types are co-ordinated only by being independently synchronised at some early stage of development to the same external timing signal. Their shared time scale being maintained through all of their differentiation processes. An alternative explanation would be that co-ordination was achieved by some sort of signalling system 'reporting' the end of meiosis to spatially distant parts of the fruit body. The route such a signal might take is not clear, but since primary gills are attached to the stem, with their central (tramal) regions in full hyphal continuity with the central stem and cap context (shown by Reijnders, 1963, 1979; Moore, 1987 and diagrammed in Fig. 1), the connection between tissues undergoing meiosis and the upper (most reactive) regions of the stem may allow direct transmission of signals for co-ordination of development in every part of the fruit body.

Table 4. *Sectional areas (μm^2) of hymenial cells measured in longitudinal sections together with pooled data for stem cells from the same specimens*

	Fruit body height (mm)							
	3	6	8	25	27	48	55	83
Stem apex	148 ± 7			292 ± 25		3184 ± 174	9243 ± 548	9109 ± 390
Stem middle			211 ± 13	3857 ± 194		6274 ± 214	10009 ± 434	10138 ± 308
Stem base				2705 ± 181		3449 ± 200	6533 ± 474	3733 ± 182
Veil cells	238 ± 14	242 ± 11	276 ± 16					
Basidia					151 ± 3	181 ± 3	177 ± 3	138 ± 3
Paraphyses					193 ± 7	244 ± 9	253 ± 7	215 ± 7
Spores					39 ± 1	48 ± 1	43 ± 1	45 ± 1
Cystidia					1194 ± 28	1423 ± 44	2495 ± 93	1391 ± 42
Cystesia					305 ± 7	303 ± 11	387 ± 17	260 ± 8

Each entry for the hymenial cell types is a mean of 50 measurements. The data for the stem cells are means of 50 to 400 measurements.

Cellular structure and patterning in the stem of Coprinus

Although the genus *Coprinus* has been one of the most keenly studied, the structure of the stem has been essentially ignored. The *Coprinus* stem is commonly described as being composed of greatly inflated and elongated cells and little else. Although Gooday (1975) mentioned the presence of narrow hyphae, the most comprehensive description of stem structure in *C. cinereus* is: 'The stem includes a central column of dikaryotic hyphae and a cortex of giant multinucleate cells' (Lu, 1974). This observation refers to a pre-meiotic fruit body and implies that the cortex is made up exclusively of inflated cells, but this is not so. The cortex comprises both narrow hyphae and inflated cells but although the cortical narrow hyphae may have been observed before no record of them appears in the literature and they tend to be dismissed as fragments of the generative (undifferentiated) hyphae which constituted the young primordium (A. F. M. Reijnders, personal communication).

Hammad, Watling & Moore (1993b) carried out a detailed morphometric analysis, counting and sizing cell profiles in transverse sections of the stem tissue using computer-aided image analysis of light microscope images. They demonstrated that the stem contains both narrow and inflated hyphae. Narrow hyphae (cross-sectional area $< 20 \ \mu m^2$) always comprise a significant numerical proportion (23% to 54%) of the cells seen in microscope sections of the stem tissue, although they only contribute 1% to 4% to the overall cross-sectional area of the stem (Table 5).

Overview of stem structure

Low magnification images of transverse sections of stems of any fruit body more than a few mm tall are dominated by the profiles of highly inflated cells (Hammad *et al.*, 1993b) though even in these images a scattering of very much narrower hyphal profiles is evident and narrow hyphae are clearly visible in longitudinal sections and scanning electron microscope images. Many histological dyes stained narrow hyphae selectively, but all staining reactions were differential in the senses that (i) only some of the narrow hyphae were stained in any one transverse section, and (ii) adjacent compartments of the same hyphal filament in longitudinal sections sometimes stained differently (for photomicrographs refer to Figs 6-9 in Hammad *et al.*, 1993b).

Table 5. *Comparison of the numbers of narrow hyphae and the area they contributed to the total area of cells in the transect for fruit bodies of* Coprinus cinereus *at different stages of development*

Stem Length (mm)	Stage of development	Narrow hyphae (% of total hyphae)	% area contributed by narrow hyphae
6	Pre-karyogamy	47.2	4.2
27	Sporulation in progress	43.6	3.4
45	Sporulation completed	35.2	2.2
70	Sporulation completed, elongation in progress	33.2	1.9

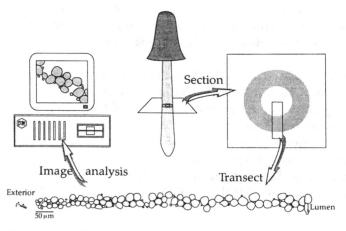

Fig. 4. Determination of cell population distributions. The diagram shows how glycolmethacrylate sections of the stem were used for image analysis of cell cross sectional areas. Transects were routinely 12 μm wide; a wider transect is shown here for illustrative convenience. (Redrawn after Hammad *et al.*, 1993b).

Size spectrum of stem cells in transverse sections

Measurements of the cross-sectional areas of hyphal profiles in 5 μm thick sections were made in transects of transverse sections of stems of various ages. Fig. 4 shows a typical transect and Figs 5 and 6 the graphical plots derived from it to analyse the spatial distributions of stem cells. All of the transects showed the sort of distribution represented in Figs 5 and 6, namely hyphae in the 0–10 and 10–20 μm^2 categories made up the largest classes of the hyphal population. Inflated hyphae had a very dispersed

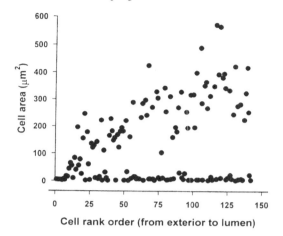

Fig. 5. Plot comparing cell cross sectional area with position of the cell in the transect (= rank order) for the transect diagrammed in Fig. 4. (From Hammad *et al.*, 1993b).

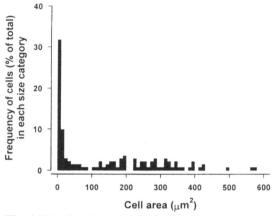

Fig. 6. Plot showing the cell size distribution of the transect diagrammed in Fig. 4. (From Hammad *et al.*, 1993b).

distribution in terms of cell area, no particular cell size being predominant. Very obviously there are two distinct populations of hyphae: narrow hyphae with cross-sectional area less than 20 μm^2, and inflated hyphae of cross-sectional area greater than or equal to 20 μm^2 (Table 5).

No single function can be assigned to narrow hyphae, all available evidence points to diverse functions. Narrow hyphae tend to be particularly concentrated at the exterior of the stem and as a lining to the lumen. On the outer surface they may serve as an insulating layer, like an epidermal layer of hairs. At the lumen they may excrete material into the cavity [Cox & Niederpruem (1975) referred to a brown gel in the lumen which disappeared as the stem extended] or merely represent the remnants of the initially central core of dikaryotic hyphae.

Narrow hyphae stained densely with a number of stains but not all narrow hyphal profiles in a transverse section and not all hyphal compartments belonging to any one narrow hypha in longitudinal sections stained equally. The reason for this differential selectivity of staining is not known, but it might reflect differential function within the population of narrow hyphae. Narrow hyphae may be important in translocation of nutrients through the stem, so differential staining may simply reflect inhomogeneities in the distribution of cytoplasmic materials during translocation. The narrow hyphae seem to form networks independent of the inflated hyphae. Narrow hyphae were branched and interconnected laterally with other narrow hyphae, but inflated hyphae were neither branched nor associated in networks. Thus narrow hyphae may be important in translocation both longitudinally (as for the supply of nutrients to the cap) and transversely (as for communication and co-ordination across the radius of the stem).

Spatial distribution of stem cells in transverse sections

Narrow hyphae were particularly evident as an outer coating of the stem and lining the lumen but they were also interspersed throughout the rest of the tissue. In sections where there was no lumen present in the stem the central region (presumptive lumen) was occupied exclusively by narrow hyphae. This was true both for extremely young fruit bodies (e.g. 6 mm tall) in which the lumen had not yet developed, and for the extreme apex of mature fruit bodies (e.g. 70 mm tall) in a region above a well developed lumen.

Statistical analysis showed that the distribution of inflated hyphae departed significantly from a non-random distribution, and tending towards evenness regardless of the age of the fruit body or position of the section within the stem. On the other hand, the spatial distribution of narrow hyphae differed significantly from randomness in only upper middle and upper apical regions of 27 mm tall and 70 mm tall fruit bodies, where there were slight tendencies towards even distributions.

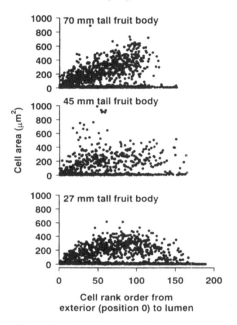

Cell area (μm^2)

Cell rank order from
exterior (position 0) to lumen

Fig. 7. Cumulated rank order plots of 27, 45 & 70 mm stems. Note how the cell size distribution across the radius of the stem changes with increasing stem length. Sizes (areas) of individual cell profiles in radial transects of transverse sections are collected in the scatter plots for each size of fruit body (1147 measurements for the 27 mm stem, 617 measurements for the 45, 1147 measurements for the 70 mm stem). (From Hammad *et al.*, 1993b).

Changes in the pattern of hyphal distribution

Narrow hyphae were interspersed with inflated hyphae across the full radius of all stems at all positions along the length of the stem and irrespective of the developmental age of the stem. However, there was a progressive change in the distribution of inflated hyphae (Fig. 7).

In 6 mm and 27 mm tall fruit bodies the inflated hyphae increased in cross-sectional area up to a point halfway across the cortex but their size declined again towards the lumen so that cells of the greatest cross-sectional area were found in the middle of the cortex. In a 45 mm tall fruit body the cross-sectional area of inflated hyphae increased gradually from the exterior to the lumen. In the 70 mm tall fruit body this pattern of increase in size right up to the margin of the lumen was even more pronounced, the peak cell area being adjacent to the lumen rather than in the mid-cortex (Fig. 7).

These observations give a very dynamic view of the way the internal cellular structure of the stem changes during its development. The measurements demonstrate that expansion of the stem involves an initial increase in size of inflated hyphae in the mid-cortex. During further development of the fruit body the cells between this zone and the lumen show the greatest expansion. Inflated hyphae around the periphery of the stem do not enlarge much. Importantly, while these changes are going on, narrow hyphae are present at all positions in the fruit body (from base to apex) and at all stages of development. The population of narrow hyphae was reduced by about 25% as size increased from 27 to 70 mm, presumably due to narrow hyphae (approx. 25%) becoming inflated. This fraction might therefore be considered to be that fragment of the primordial generative hyphae preserved as a reserve of hyphal inflation capacity to support final maturation growth; but it is only a minority fraction of the narrow hyphal population. Other members of that population have other functions. If inflated hyphae do arise by expansion of the randomly distributed narrow hyphae, the even (i.e. non-random) distribution of inflated hyphae implies that a pattern forming process determines their differentiation.

The mechanical consequences of this pattern of cell inflation are simple. Increase in cross-sectional area of inflated hyphae in the middle of the cortex will (i) result in the central core being torn apart, leaving its constituent cells as a remnant around the inner wall of the lumen so created; and (ii) stretch, reorganise and compress the tissues in the outer zones of the stem (Fig. 8).

Thus, formation of the mature stem as a cylinder with outer tissues under tension and inner tissues in compression (the optimum mechanical structure for a vertical cylindrical support) is entirely a result of the pattern of cell inflation within the stem as the stem develops. This specific pattern of inflation must be organised by signalling molecules which determine differential cell inflation across the stem radius.

What makes some of the hyphae differentiate into inflated and multinucleate structures while the narrow hyphae remain morphologically similar to the vegetative mycelial hyphae is not known. This differentiation occurs at an extremely early stage as both narrow and inflated hyphae were seen in 3 mm tall primordia.

Cell distributions in other fungal fruit bodies

The studies of Hammad et al. (1993a, b) on Coprinus cinereus remain the only thorough quantitative analyses of cell distributions in any fungal fruit body. All other studies in the literature are traditional observational ones.

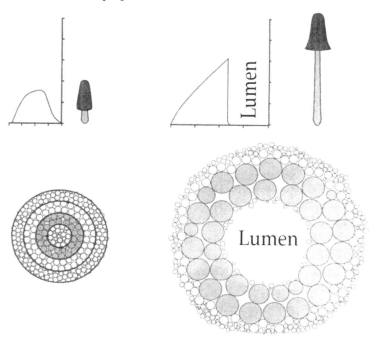

Fig. 8. Interpretation of the geometrical consequences of the cell size changes during development revealed by the data of Fig. 7. The graphs in the upper part of the figure show the lines of best fit for the scatter plots (Fig. 7) of 27 and 70 mm stems, together with scale drawings of the fruit bodies. The diagrams in the lower part of the figure are transverse sections of the stems, drawn to scale. The diagrammatic transverse section of the stem of a 27 mm tall primordium is (on left) is composed of solid tissue which is divided into four zones corresponding to the zones in the radius in the graph above. The central (zone 4) and outermost (zone 1) zones are comprised of rather smaller cells than the two cortical zones. During further growth the most dramatic cell inflation occurs in the cells of zone 3 which are here shown shaded. Growth from 27 mm to 70 mm in height is accompanied by a 3.6-fold increase in cell area in zone 3 and a 1.6-fold increase in zone 2. The cells in the other zones have to be rearranged to accommodate the inflation of zone 3 and a major consequence is that a lumen appears in the centre of the stem. Changes in area are shown accurately to scale in these diagrams, though only a total of 327 cell profiles are illustrated. As this is only a tiny proportion of the cells involved *in vivo*, the diagrams inevitably distort the apparent relationship between cell size and stem size. Narrow hyphae have been ignored in these diagrams, though they are distributed throughout the tissue and their conversion to inflated hyphae contributes to stem expansion.

Corner (1932) introduced hyphal analysis as a procedure to encompass descriptive studies of hyphal systems in polypore fruit bodies (for review see Pegler, 1996). Corner introduced the terms monomitic, dimitic and trimitic to describe tissues consisting of one, two or three kinds of hyphae, and hyphae in these different categories have been referred to as generative (because they ultimately give rise to the basidia and directly or indirectly to all other structures), skeletal (thick walled with narrow lumen, but lacking branching and septation) or binding hyphae (which have limited growth and irregular often repeated branching). Corner (1966) later coined the terms sarcodimitic and sarcotrimitic to describe particular hyphal systems where there are two or three types of hyphae of which one is inflated and thickened. Redhead (1987) was able to recognise a whole group of closely related agarics with such structures but this included neither the Coprinaceae nor the Russulaceae. Fayod (1889) had already demonstrated that the trama in Russulaceae consists of a mixture of swollen cells (sphaerocysts) and filamentous hyphae, which he called 'fundamental hyphae', but it was Reijnders (1976) who showed that there is a developmental pattern in the tissue with some of the filamentous hyphae playing a very important role. He has described aggregations of hyphae, termed hyphal knots (Reijnders, 1977) in a wide range of species (Reijnders, 1993). The common features of Reijnders' hyphal knots seem to be a central hypha (which remains hyphal) and an immediately-surrounding family of hyphae which differentiate in concert. Such structures have also been observed in *Lactarius* (Watling & Nicoll, 1980). Thus, swollen cells in a ring or cylinder around a central hypha may be formed by many species (another example is described in Section 7.5) and may be the basis of longer range positional control of patterns (see Section 7.7).

The cystidial morphogenetic field

The hymenium of *Coprinus cinereus* contains four cell types: basidia, paraphyses, cystidia and cystesia. The paraphyses expand to become the major structural members of the gill lamellae, but they arise secondarily as branches from sub-basidial compartments and when first formed, the hymenium consists of a carpet of probasidia with a scattering of cystidia (Rosin & Moore, 1985a). According to Smith (1966), cystidia '. . . occur haphazardly in the hymenium, depending on the species, and vary from abundant to absent . . .'; a description of a cell distribution pattern which, in terms of developmental biology is totally useless. Much of the literature dealing with cystidia is similarly blinkered, considering them from the taxonomic viewpoint (e.g. Lentz, 1954; Price, 1973), reflecting assiduous attention to fine distinctions of nomenclature at the expense of appreciation of the remarkable developmental plasticity revealed by their occurrence and form. Brefeld (1887, cited in Buller, 1910) concluded that cystidia are metamorphosed basidia, a view summarised by Corner (1947) in the phrase '...cystidia represent sterile basidia which become overgrown...' Certainly, young basidia and young cystidia both originate as the terminal compartments of branches from the hyphae of the sub-hymenium, but cystidia are NOT overgrown basidia. The mature cystidium is a cell that is highly differentiated for its particular function (see below).

Fuzzy logic in fungal differentiation

Development of a cystidium represents expression of a perfectly respectable pathway of differentiation and commitment of a hyphal tip to the cystidial as opposed to the basidial pathway of differentiation. The commitment must occur very early in development of the hymenium because young cystidia are recognisable in the very earliest stages (Rosin & Moore, 1985a, b). The controls which determine formation of a cystidium, instead of a basidium, by a particular hyphal apex need to be established. It is certainly the case that the basidial developmental pathway (in *Agaricus bisporus*) can be interrupted to allow this cell type to serve a structural rather than spore-producing function (Allen, Moore & Elliott, 1992), though this is clearly arrested meiosis, not sterility. Similarly, it is also evident that cells which are undoubtedly cystidia on morphological criteria can occasionally show evidence of entry into meiosis (Chiu & Moore, 1993; and see Chapter 5), which suggests that entry to the cystidial pathway of differentiation does not totally preclude expression of at least part of the meiocyte differentiation pathway. Similarly, the fact that a large fraction of the *in situ* basidial population of *A. bisporus* remains in arrested meiosis indicates that entry to the meiotic division pathway does not guarantee sporulation; a fact also demonstrated with excised gills of *C. cinereus in vitro* (Chiu & Moore, 1990a).

Further examples can be found in the literature. Watling (1971) observed some cystidia bearing hyphal outgrowths looking like sterigmata in a spontaneous fruit body variant of *Psilocybe merdaria*, while Schwalb (1978) reported that basidia of a temperature-sensitive mutant of *Schizophyllum commune* not only aborted meiosis but also produced elongated sterigmata at the restrictive temperature. A spore-deficient mutant of *Lentinula edodes* (= *Lentinus edodes*) produced some abnormal basidia bearing both a hyphal outgrowth and basidiospores (Hasebe, Murakami & Tsuneda, 1991).

Similar abnormalities in basidia have been induced in *Coprinus cinereus* by transplanting gills to agar medium containing some metabolic inhibitors (Chiu & Moore, 1988a & b; 1990a; and see Chapter 6). These explantation experiments have been discussed mainly for their value in understanding commitment to the basidium differentiation pathway, but it is equally important that all other cells of the hymenium and hymenophore showed no commitment; immediately reverting to hyphal growth on explantation. This implies that all differentiated cells except the meiocyte have an extremely tenuous grasp on their state of differentiation; so tenuous that when removed from their normal tissue environment they revert immediately to the vegetative hyphal growth mode. That these cells do not default to hyphal growth *in situ* implies that their state of differentiation is somehow continually reinforced by some aspect of the environment of the tissue which they comprise. Interestingly, although cystidia of *Coprinus* reverted to hyphal growth when excised, cystidia of excised gills of *Volvariella bombycina* were arrested and did not show reversion to vegetative growth suggesting they are another differentiated hymenial cell type (Chiu & Moore, unpublished).

Chiu & Moore (1993) discuss the possibility that fungal differentiation pathways exhibit what would be described as 'fuzzy logic' in cybernetic terms. Instead of viewing fungal cell differentiation as involving individual major 'decisions' which switch progress between alternative developmental pathways which lead inevitably to specific combinations of features, this idea suggests that the end point in fungal differentiation depends on the balance of a number of minor 'decisions'. So, rather than rigidly following a prescribed sequence of steps, fungal differentiation pathways are based on application of rules which allow considerable latitude in expression. Developmental decisions between pathways of differentiation seem to be able to cope with a degree of uncertainty, allowing fungal cells to assume a differentiation state even when all conditions of that state have not been met.

Distribution of cystidia

Descriptions of cystidium distribution commonly encountered in the literature feature adjectives like 'scattered', 'haphazard', 'fairly uniform'. The only statistically valid description of cystidial distribution has been published by Horner & Moore (1987).

Cystidia are found 'scattered' in fair number over the hymenium of *C. cinereus*. Visually, cystidial density-distribution on the face of the gill is fairly uniform but at the gill edge the density of cystidia is locally increased. When sufficiently developed, cystidia span the gill cavity, their apices adhering to cystesia in the opposite hymenium. At early stages in growth of

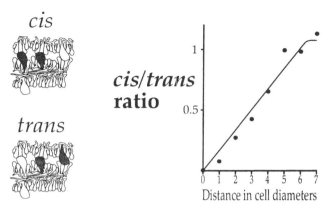

Fig. 9. Cystidium distribution in the *Coprinus cinereus* hymenium. The drawings on the left of the graph show the categorisation of neighbouring pairs of cystidia in micrographs as either *cis* (both emerge from the same hymenium) or *trans* (emerging from opposite hymenia). The plot compares the frequencies of these two types over various distances of separation and shows that closely-spaced *cis* neighbours are less frequent than closely-spaced *trans* neighbours, implying some inhibitory influence over the patterning of cystidia emerging from the same hymenium. (Redrawn after Moore, 1995).

the cystidium the cell(s) with which it will eventually come into contact in the opposing hymenium are indistinguishable from their fellow probasidia. However, when the cystidium comes firmly into contact with the opposing hymenium, the hymenial cells with which it collides develop a distinct granular and vacuolated cytoplasm, more akin to that of the cystidium itself than to the neighbouring probasidia. This suggests that a contact stimulus sets in train an alternative pathway of differentiation leading to an adhesive cell type called the cystesium.

About 8% of the hyphal tips in the protohymenium of *C. cinereus* become cystidia (Rosin & Moore, 1985b). Cystidia may arise from either of the two hymenia which are on opposite sides of each gill cavity and they can be seen, counted and measured easily in sections cut for light microscope observation (details in Horner & Moore, 1987). Demonstration of the cystidial morphogenetic field relied upon measurements made of such sections. The argument was based on the expectations that cystidia spanning the gill cavity may be 'distant', having other cells separating them, or 'adjacent', with no intervening cells; and, in either case, both cystidia may emerge from the same hymenium (described as 'cis') or from opposite hymenia ('trans'). *A priori*, one would predict that if the distribution of cystidia is entirely randomised then the frequency of adjacent pairs will

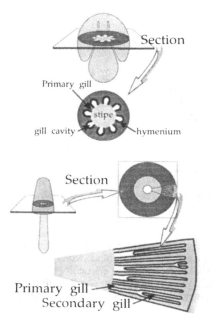

Fig. 10. Diagrammatic cross section of primordium (top) and older fruit body (bottom) showing primary and secondary gills. (Redrawn after Moore, 1995).

depend on the population density only and there will be an equal number of cis and trans in both the distant and adjacent categories. However. quantitative data from serial sections showed that there was a positive inhibition of formation of neighbouring cystidia in the same hymenium (Fig. 9).

The interpretation of data shown in Fig. 9 is that formation of a cystidium actively lowers the probability of another being formed in the immediate vicinity. The extent of the inhibitory influence extends over a radius of about 30 μm and is strictly limited to the hymenium of origin. For the activating component of the system it could be suggested that differentiation leading to cystidium formation is activated by the concentration of a constituent of the atmosphere in the gill cavity immediately above the developing hymenium. Many cystidia seem to be secretory (Chiu & Moore, 1990c) so their metabolism may well react to the local gaseous environment. The distribution pattern of cystidia is thus interpreted as being dependent on interplay between activating and inhibiting factors which define the cystidial morphogenetic field.

Forming gills into patterns

Most of the patterns discussed so far are particular cell distributions which are evident only on the microscopic scale. The most obvious pattern in an agaric fruit body, though, is the distribution of the gills. The gills are essentially tissue plates suspended from the underside of the fruit body cap. In *Coprinus*, the cap of the fruit body primordium encloses the top of the stem and gills are formed as vertical plates arranged radially around the stem (Fig. 10).

There are two types of gill (diagrammed in Fig. 10): primary gills which, from formation, have their inner (tramal) tissue in continuity with the outer layers of the stem, and secondary (and lesser ranked) gills in which the hymenium is continuous over the gill edge (Reijnders, 1979; Rosin & Moore, 1985a; Rosin *et al.*, 1985; Moore, 1987).

The direction of growth

Crucial to a proper understanding of the morphogenesis of fruit bodies is knowledge of the direction of growth of the constituent parts. Agaric gills are suspended from the flesh of the cap and the intuitive expectation is that gills extend at their free edge. This direction of growth (i.e. radially inwards towards the stem) seems to be assumed as an unassailable truism in most of the earlier literature. Logic seems to be in its favour. After all, fungal hyphae characteristically show apical growth, and the free edge of the gill is composed of hyphal apices; should they not extend the gill?

Unfortunately, this thought immediately raises a problem. The hyphal tips which make up the free edge of the gill are terminally differentiated into hymenial cells; they have lost their capacity for extension growth. Primary gills of *Coprinus* spp. present another problem because they are connected with cap tissue at their outer edge and with the stem at their inner edge. How does a gill without a free edge extend? The observations and experiments discussed below show that while it is not unreasonable nor illogical to assume that the direction of growth is radially inwards towards the stem, it is, in fact, diametrically wrong.

During gill development in *Volvariella bombycina* it has been demonstrated that the free edges of the gills remain essentially intact. This was done by painting black ink marks on the tissues in a primordium (Chiu & Moore, 1990a). During further fruit body development, ink marks placed on the cap margin and those placed on the edges of the gills remained at the margin or the gill edges respectively. The growth increment in these

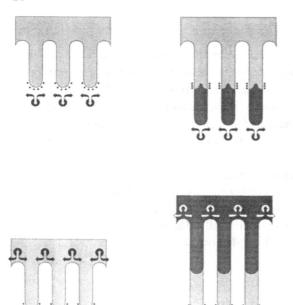

Fig. 11. Gill formation in *Volvariella bombycina*. Line drawings illustrating the outcome of marking experiments (Chiu & Moore, 1990a) and what they reveal about the direction of gill development in this organism. The diagrams show diagrammatic sections of primordial gills on the margins of which ink particles have been placed (diagrams on the left). The diagrams on the right illustrate alternative outcomes from further growth depending on whether growth occurs at the gill margin (gill organiser located at the gill margin; top right) or at the gill base (gill organiser located at the foot of the gill; bottom right). In the experiments reported by Chiu & Moore (1990a), ink marks painted on the gill margins of primordial fruit bodies were still clearly visible on the gill margins of mature fruit bodies, demonstrating that agaric gills grow by extension at their roots, and not by extension from the free margin. (Redrawn after Moore, 1995).

experiments was quite considerable, the radius of the cap increasing from 0.5 to 2.5 cm and the depth of the gills from 1.5 to 5 mm. If growth of the cap and gill margins resulted from apical growth of the hyphal tips which occupied the margin, then ink particles placed on those hyphal tips would be left behind as the hyphal apices extended which would consequently have resulted in the ink marks being buried beneath 4 to 20 mm of newly formed tissue by the end of the experiment (Fig. 11). It follows, therefore, that the hyphal tips which first form the cap margin, and those which form the gill margin, always remain at the margin. They do not continue to grow apically to extend the margin radially, nor are they overtaken by other

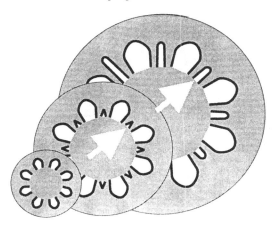

Fig. 12. Diagrammatic transverse sections of developing primordia of *Coprinus* illustrating that because primary gills are attached to the stem, their thickness must increase as the stem circumference increases. To avoid formation of excessively thick gills, new gill cavities must arise within the central (tramal) tissue of primary gills (arrows). Their progression outwards identifies the prime direction of development in this fruit body. (Adapted and redrawn after Moore, 1995).

hyphae; instead they are 'pushed' radially outwards by the press of fresh growth behind, and they are joined by fresh branches appearing alongside as the circumference of the margin is increased (Fig. 11).

There seems to be an exactly homologous process in *Coprinus*. Gills in young *Coprinus* primordia are not open to easy manipulation, so in this case the argument depends upon observation of particular gill structures during development. The crucial observation derives from the fact that primary gills of *Coprinus* spp. are connected with cap tissue at their outer edge and with the stem at their inner edge combined with consideration of the geometrical consequences of this arrangement as the tissues expand (as depicted in Fig. 1). The gills are attached to the stem circumference, but the stem circumference must increase greatly during development. If the gills are attached to a stem of increasing circumference, why does gill thickness at the point of attachment not increase in proportion? The tendency to widen as the stem circumference increases is compensated by gill replication, and specifically by formation of a new gill cavity and its bounding pair of hymenia within the trama of a pre-existing gill (Moore, 1987). This clearly sets the direction of development as outwards from the stem; i.e. gills in the *C. cinereus* fruit body grow radially outwards, their roots extending into the undifferentiated tissue of the pileus context (Fig. 12).

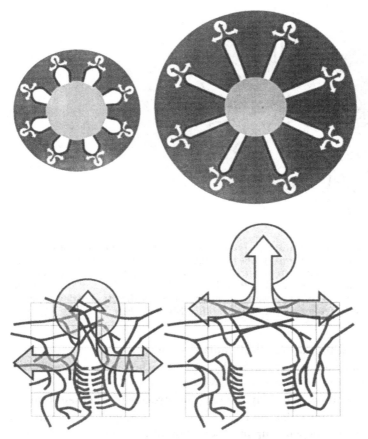

Fig. 13. The outward progression of gill cavities (upper diagrams) is managed by gill organisers which migrate radially into newly-produced undifferentiated tissue. As they pass through an undifferentiated region, gill organisers promote the branching pattern that generates the two opposing hymenia of neighbouring gills (bottom left). Subsequent expansion of the inner tissues separates the two hymenia (depicted in the 'stretched' co-ordinates of the diagram at lower right) to produce the gill cavity. (Adapted and redrawn after Moore, 1995).

Two different sources of information consequently demonstrate that growth of an agaric gill occurs by outward progression of gill tissue at its foot. Gill growth proceeds radially outwards; differentiation of gill tissue extending into the undifferentiated tissue of the outwardly expanding cap tissue (Fig. 12).

Embryonic gills are convoluted

In both *C. cinereus* and *V. bombycina* gills are formed as convoluted plates (Chiu & Moore, 1990b). A sinuous, labyrinthiform hymenophore appears to be a normal 'embryonic' stage in fruit body development in agarics, yet a regular radial arrangement of the gills is characteristic of the mature basidiome. How this is achieved is a function of the expansion of the maturing primordium generating stresses between tissue layers which stretch or inflate the convoluted gills into strict radii. Stretching is the effective force in *C. cinereus*; the cystidia being critical elements in the communication of the formative stresses around the fruit body. Inflation occurs in *V. bombycina*: the tightly appressed cells of the hymenium forming a tensile layer containing the compression generated by cell inflation in the gill trama.

The gill organizer

A fundamental 'rule' during the very earliest stages of agaric gill formation seems to be: if there is room, make a gill; without reference to the exact spatial orientation of the gill tissue so formed. The mechanics of fruit body expansion will compensate for any meandering in the direction taken during gill formation. The formative element which directs the development of undifferentiated tissue of the cap into gill tissue is an organiser in the tissue at the extreme end of the gill cavity (Fig. 13).

The gill organiser is responsible for the progression of the gill cavity radially outwards, away from the stem. It directs the 'undifferentiated-to-differentiated' transition. Presumably this is largely an increase in branch frequency to produce branches of determinate growth which are mutually 'attracted' so that they form the opposing young hymenia on either side of what will become the gill cavity. Cap expansion separates the two protohymenia, thus extending the gill cavity deeper into the tissue of the cap. Since they are progressing radially outwards, neighbouring organisers become further and further separated from one another as development proceeds and as the distance between neighbouring organisers increases a new one can arise between them (diagrammed in Fig. 13; micrographs shown in Rosin & Moore, 1985a); when a new gill organiser emerges, the margin of a new (but 'secondary') gill is formed. It is extended not by growth of its margin, but by continued radial outward progression of the two (mother and daughter) gill organisers which now straddle either side of its root.

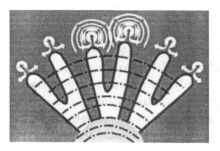

Fig. 14. Continued outward progression of gill organisers could be their response to a radial gradient within the inner tissues (which might be an outwardly diffusing signal or a metabolic concentration gradient). In addition, each gill organiser could control its own morphogenetic field by producing an inhibitor that prevents a new gill organiser arising within its diffusion range. As existing gill organisers move radially away from each other the inhibitory effect diminishes and a new organiser can be produced in response to the primary signal from the centre. (Adapted and redrawn after Moore, 1995).

In the origin of the gills we see operating two classic components of theoretical morphogenesis – activation and inhibition by diffusing morphogens. First, we can assume that diffusion of an activating signal along the fruit body radius assures progression of the gill organiser along its outward radial path. Second, each organiser can be assumed to produce an inhibitor which prevents formation of a new organizer within its diffusion range (i.e. the gill organiser uses this inhibitor to control its morphogenetic field) (Fig. 14). As radial progression into the extending (undifferentiated) cap tissue causes neighbouring organisers to diverge, a region appears between them which is beyond the range of their inhibitors – at this point a new organizer can arise in response to the radial activating signal. Interactions between the diffusion and decay characteristics of the activator and the inhibitor are all that is necessary to control gill spacing, gill number, gill thickness, and the radial orientation of the gill field.

The origin of space

The internal structure of the *Coprinus* primordium is uniformly solid at the time that gills begin to arise, so gills and gill space arise together. Similarly, as is argued above, as development proceeds gills, hymenia and the gill spaces between them emerge together as the gill organisers migrate outwards into undifferentiated cap tissue.

Lu (1991) has claimed that the gill cavities in *Coprinus* arise as a result of 'programmed cell death' (originally 'cell disintegration' (Lu, 1974)) after observing multivesicular and membranous bodies in cells of the gill cavities on chemically fixed and processed transmission electron microscope specimens. He interpreted these observations to indicate that cell disintegration accounts for formation of the gill cavity, proposing a selective and programmed cell death as a part of tissue remodelling during gill formation. The hypothesis of programmed cell death is attractive but there is not enough evidence for it. In animal cell systems, cell death (apoptosis) is an essential component of embryogenesis as well as a continuing strategy for control of the dynamic balance of tissues in the body (Sen, 1992). Apoptosis is a very specific cellular process which exhibits a number of characteristic cytological and molecular processes. None of these have been observed in fungi. Further, cell disintegration has not been reported in any earlier studies on *Coprinus* (Reijnders, 1963, 1979; Rosin & Moore, 1985a, b) although similar vesicular aggregations have been ascribed to fixation artifacts (Waters, Butler & Moore, 1975a, b). The balance of evidence, therefore, is that the case has not been made for apoptosis generating the spaces which occur in fungal fruit bodies.

Moore (1995) pointed out that cell degeneration, whether by necrosis or apoptosis, was unnecessary to create spaces within a fruit body which is continually expanding. All that is needed is that branching patterns are organised to form 'surfaces' which can separate when the fruit body expands. He called the process *cavitation* and suggested that spaces can be formed in the following way. When branches of determinate growth (which are mutually 'attracted') are formed opposing one another as a pair of palisaded cell plates (like the opposing hymenia of neighbouring gills), they form an incipient fracture plane. This plane can be opened out into a cavity when the expansion of the underlying tissue puts tension across the 'fracture' and pulls the palisades apart. The argument applies to cavitation in all differentially expanding cellular structures. Variations on the theme can be imagined in other regions and other organisms. If the 'fracture planes' form an annulus around the top of the stem (one tissue layer might be the stem apical meristemoid, the other the hymenophore meristemoid), then an annular cavity could arise before gill formation.

Conclusions

Even the most casual observations confirm that co-ordination of developmental processes is successfully achieved in fungal multicellular structures, but there is little convincing evidence for chemicals able to perform the signal communication involved (Chapter 7). Thus, there are no clues to the nature of the growth factors involved in these phenomena. Also, given that lateral contacts between fungal hyphae are rare in comparison with the lateral interconnection by plasmodesmata, gap junctions and cell processes in cells of plant and animal tissues, any morphogens responsible for coordinating the activities of different hyphal branching networks are likely to be communicated exclusively through the extracellular environment (Reijnders & Moore, 1985).

Perhaps the ultimate morphogenetic regulatory unit in multihyphal fungal structures is the Reijnders hyphal knot - a little community comprising an induction hypha (or hyphal tip, or hyphal compartment) and the immediately surrounding hyphae (or tips, or compartments) which can be brought under its influence. Larger scale morphogenesis could be co-ordinated by 'knot-to-knot' interactions (Moore, 1995).

Many of the phenomena described in this chapter are poor candidates for experiment but the chapter shows how careful application of essentially conventional but quantitative observational methods can extricate valuable information. Direct experiment requires a morphogenetic model system. This, too, seems to be available: in one case through the use of explantation experiments (Bastouill-Descollonges & Manachre, 1984; Chiu & Moore, 1988a & b, 1990a; and see Chapter 6) and also through the use of tropisms which are simple developmental pattern forming processes able to generate a particular morphogenetic change on demand in a specific location (Moore, 1991; Moore *et al.*, 1996).

Significant advances in fungal developmental biology will only be achieved if more use is made of quantitative methods. The science is doomed to stagnation if, in the days of computer-aided image analysis and virtual reality, the practitioners of mycology remain satisfied with hand drawn line art and subjective verbal description to communicate their observations. Fungal developmental biology must get into the 20th century before the 21st brushes it aside.

Acknowledgements

This chapter was written during a visit to the Chinese University of Hong Kong and I thank the Leverhulme Trust for the award of a Research Grant which enabled this, the Department of Biology at CUHK for their hospitality, and The University of Manchester for leave of absence.

References

Allen, J. J., Moore, D. & Elliott, T. J. (1992). Persistent meiotic arrest in basidia of *Agaricus bisporus*. *Mycological Research* **96**, 125–127.

Bastouill-Descollonges, Y. & Manachère, G. (1984). Photosporogenesis of *Coprinus congregatus*: correlations between the physiological age of lamellae and the development of their potential for renewed fruiting. *Physiologia Plantarum* **61**, 607–610.

Bonner, J. T., Kane, K. K. & Levey, R. H. (1956). Studies on the mechanics of growth in the common mushroom, *Agaricus campestris*. *Mycologia* **48**, 13–19.

Borriss, H. (1934). Beitrage zur Wachstums und Entwicklungsphysiologie der Fruchtkörper von *Coprinus lagopus*. *Planta* **22**, 28–69.

Bret, J. P. (1977). Respective role of cap and mycelium on stipe elongation of *Coprinus congregatus*. *Transactions of the British Mycological Society* **68**, 262–269.

Buller, A. H. R. (1910). The function and fate of the cystidia of *Coprinus atramentarius*, together with some general remarks on *Coprinus* fruit bodies. *Annals of Botany* **24** (old series), 613–629.

Buller, A. H. R. (1924). *Researches on Fungi*, vol. 3. Longman Green & Co.: London.

Chiu, S. W. & Moore, D. (1988a). Evidence for developmental commitment in the differentiating fruit body of *Coprinus cinereus*. *Transactions of the British Mycological Society* **90**, 247–253.

Chiu, S. W. & Moore, D. (1988b). Ammonium ions and glutamine inhibit sporulation of *Coprinus cinereus* basidia assayed *in vitro*. *Cell Biology International Reports* **12**, 519–526.

Chiu, S. W. & Moore, D. (1990a). Sporulation in *Coprinus cinereus*: use of an *in vitro* assay to establish the major landmarks in differentiation. *Mycological Research* **94**, 249–259.

Chiu, S. W. & Moore, D. (1990b). A mechanism for gill pattern formation in *Coprinus cinereus*. *Mycological Research* **94**, 320–326.

Chiu, S. W. & Moore, D. (1990c). Development of the basidiome of *Volvariella bombycina*. *Mycological Research* **94**, 327–337.

Chiu, S. W. & Moore, D. (1993). Cell form, function and lineage in the hymenia of *Coprinus cinereus* and *Volvariella bombycina*. *Mycological Research* **97**, 221–226.

Corner, E. J. H. (1932). A *Fomes* with two systems of hyphae. *Transactions of the British Mycological Society* **17**, 51–81.

Corner, E. J. H. (1947). Variation in the size and shape of spores, basidia and cystidia in Basidiomycetes. *New Phytologist* **46**, 195–228.

Corner, E. J. H. (1966). *A monograph of cantharelloid fungi*. Annals of Botany Memoirs no. 2. Oxford University Press: London.

Cox, R. J. & Niederpruem, D. J. (1975). Differentiation in *Coprinus lagopus*. III. Expansion of excised fruit bodies. *Archives of Microbiology* **105**, 257–260.

Craig, G. D., Gull, K. & Wood, D. A. (1977). Stipe elongation in *Agaricus bisporus*. *Journal of General Microbiology* **102**, 337–347.

Eilers, F. I. (1974). Growth regulation in *Coprinus radiatus*. *Archives of Microbiology* **96**, 353–364.

Fayod, V. (1889). Prodrome d'une histoire naturelle des Agaricinés. *Annales des Sciences Naturelles, Botanique Série* **7–9**, 179–411.

Gooday, G. W. (1974). Control of development of excised fruit bodies and stipes of *Coprinus cinereus. Transactions of the British Mycological Society* **62**, 391–399.

Gooday, G. W. (1975). The control of differentiation in fruit bodies of *Coprinus cinereus. Reports of the Tottori Mycological Institute (Japan)* **12**, 151–160.

Gooday, G. W. (1982). Metabolic control of fruit body morphogenesis in *Coprinus cinereus.* In *Basidium and Basidiocarp: Evolution, Cytology, Function and Development* (ed. K. Wells & E. K. Wells), pp. 157–173. Springer–Verlag: New York.

Gooday, G. W. (1985). Elongation of the stipe of *Coprinus cinereus.* In *Developmental Biology of Higher Fungi* (ed. D. Moore, L. A. Casselton, D. A. Wood & J. C. Frankland), pp. 311–331. Cambridge University Press: Cambridge, U.K.

Gruen, H. (1982). Control of stipe elongation by the pileus and mycelium in fruitbodies of *Flammulina velutipes* and other Agaricales. In *Basidium and Basidiocarp: Evolution, Cytology, Function and Development* (ed. K. Wells & E. K. Wells), pp. 125–155. Springer–Verlag: New York.

Gruen, H. (1991). Effects of grafting on stipe elongation and pileus expansion in the mushroom *Flammulina velutipes. Mycologia* **83**, 480–491.

Hafner, L. & Thielke, C. (1970). Kernzahl und Zellgrösse im Fruchtkorperstiel von *Coprinus radiatus* (Solt.) Fr. *Berichte Deutsche Botanische Gesellschaft* **83**, 27–31.

Hammad, F., Ji, J., Watling, R. & Moore, D. (1993a). Cell population dynamics in *Coprinus cinereus*: co-ordination of cell inflation throughout the maturing basidiome. *Mycological Research* **97**, 269–274.

Hammad, F., Watling, R. & Moore, D. (1993b). Cell population dynamics in *Coprinus cinereus*: narrow and inflated hyphae in the basidiome stipe. *Mycological Research* **97**, 269–274.

Hasebe, K., Murakami, S. & Tsuneda, A. (1991). Cytology and genetics of a sporeless mutant of *Lentinus edodes. Mycologia* **83,** 354–359.

Horner, J. & Moore, D. (1987). Cystidial morphogenetic field in the hymenium of *Coprinus cinereus. Transactions of the British Mycological Society* **88**, 479–488.

Kamada, T., Kurita, R. & Takemaru, T. (1978). Effects of light on basidiocarp maturation in *Coprinus macrorhizus. Plant & Cell Physiology* **19**, 263–275.

Kamada, T., Miyazaki, S. & Takemaru, T. (1976). Quantitative changes of DNA, RNA and protein during basidiocarp maturation in *Coprinus macrorhizus. Transactions of the Mycological Society of Japan* **17**, 451–460.

Kamada, T. & Takemaru, T. (1977). Stipe elongation during basidiocarp maturation in *Coprinus macrorhizus*: mechanical properties of stipe cell wall. *Plant & Cell Physiology* **18**, 831–840.

Lentz, P. L. (1954). Modified hyphae of hymenomycetes. *The Botanical Review* **20**, 135–199.

Lu, B. C. (1974). Meiosis in *Coprinus.* V. The role of light on basidiocarp initiation, meiosis, and hymenium differentiation in *Coprinus lagopus. Canadian Journal of Botany* **52**, 299–305.

Lu, B. C. (1991). Cell degeneration and gill remodelling during basidiocarp development in the fungus *Coprinus cinereus. Canadian Journal of Botany* **69**, 1161–1169.

Madelin, M. F. (1956). Studies on the nutrition of *Coprinus lagopus* Fr. especially as affecting fruiting. *Annals of Botany* **20**, 307–330.

Matthews, T. R. & Niederpruem, D. J. (1973). Differentiation in *Coprinus lagopus*. II. Histology and ultrastructural aspects of developing primordia. *Archives für Mikrobiologie* **88**, 169–180.

McLaughlin, D. J. (1982). Ultrastructure and cytochemistry of basidial and basidiospore development. In *Basidium and Basidiocarp: Evolution, Cytology, Function and Development* (ed. K. Wells & E. K. Wells), pp. 37–74. Springer–Verlag: New York.

Moore, D. (1987). The formation of agaric gills. *Transactions of the British Mycological Society* **89**, 105–108.

Moore, D. (1991). Perception and response to gravity in higher fungi – a critical appraisal. *New Phytologist* **117**, 3–23.

Moore, D. (1995). Tissue formation. In *The Growing Fungus* (ed. N. A. R. Gow & G. M. Gadd), pp. 423–465. Chapman & Hall: London.

Moore, D., Elhiti, M. M. Y. & Butler, R. D. (1979). Morphogenesis of the carpophore of *Coprinus cinereus*. *New Phytologist* **83**, 695–722.

Moore, D. & Ewaze, J. O. (1976). Activities of some enzymes involved in metabolism of carbohydrate during sporophore development in *Coprinus cinereus*. *Journal of General Microbiology* **97**, 313–322.

Moore, D., Hock, B., Greening, J. P., Kern, V. D., Novak Frazer, L. & Monzer, J. (1996). Centenary review. Gravimorphogenesis in agarics. *Mycological Research* **100**, 257–273.

Morimoto, N. & Oda, Y. (1973). Effects of light on fruit-body formation in a basidiomycete, *Coprinus macrorhizus*. *Plant & Cell Physiology* **14**, 217–225.

Morimoto, N. & Oda, Y. (1974). Photo-induced karyogamy in a basidiomycete, *Coprinus macrorhizus*. *Plant & Cell Physiology* **15**, 183–186.

Pegler, D. N. (1996). Centenary Review: Hyphal analysis of basidiomata. *Mycological Research* **100**, 129–142.

Price, I. P. (1973). A study of cystidia in effused Aphyllophorales. *Nova Hedwigia* **24**, 515–618.

Pukkila, P. J. & Lu, B. C. (1985). Silver staining of meiotic chromosomes in the fungus, *Coprinus cinereus*. *Chromosoma* **91**, 108–112.

Pukkila, P. J., Yashar, B. M. & Binninger, D. M. (1984). Analysis of meiotic development in *Coprinus cinereus*. In *Controlling Events in Meiosis* (ed. C. W. Evans & H. G. Dickinson), Symposia of the Society for Experimental Biology **38**, pp. 177–194. Cambridge University Press: Cambridge, U.K.

Redhead, S. A. (1987). The Xerulaceae (Basidiomycetes): a family with sarcodimitic tissues. *Canadian Journal of Botany* **65**, 1551–1562.

Reijnders, A. F. M. (1948). Études sur le développement et l'organisation histologique des carpophores dans les Agaricales. *Recuil des Travaux Botaniques Nerlandais* **41**, 213–396.

Reijnders, A. F. M. (1963). *Les problèmes du développement des carpophores des Agaricales et de quelques groupes voisins*. Dr W. Junk: The Hague.

Reijnders, A. F. M. (1976). Recherches sur le développement et l'histogénèse dans les Asterosporales. *Persoonia* **9**, 65–83.

Reijnders, A. F. M. (1977). The histogenesis of bulb and trama tissue of the higher Basidiomycetes and its phylogenetic implications. *Persoonia* **9**, 329–362.

Reijnders, A. F. M. (1979). Developmental anatomy of *Coprinus*. *Persoonia* **10**, 383–424.

Reijnders, A. F. M. (1993). On the origin of specialised trama types in the Agaricales. *Mycological Research* **97**, 257–268.

Reijnders, A. F. M. & Moore, D. (1985). Developmental biology of agarics – an overview. In *Developmental Biology of Higher Fungi* (ed. D. Moore, L. A. Casselton, D. A. Wood & J. C. Frankland), pp. 333–351. Cambridge University Press: Cambridge, U.K.

Rosin, I. V., Horner, J. & Moore, D. (1985). Differentiation and pattern formation in the fruit body cap of *Coprinus cinereus*. In: *Developmental Biology of Higher Fungi* (ed. D. Moore, L. A. Casselton, D. A. Wood & J. C. Frankland), pp. 333–351. Cambridge University Press: Cambridge, U.K.

Rosin, I. V. & Moore, D. (1985a). Origin of the hymenophore and establishment of major tissue domains during fruit body development in *Coprinus cinereus*. *Transactions of the British Mycological Society* **84**, 609–619.

Rosin, I. V. & Moore, D. (1985b). Differentiation of the hymenium in *Coprinus cinereus*. *Transactions of the British Mycological Society* **84**, 621–628.

Schwalb, M. N. (1978). Regulation of fruiting. In *Genetics and Morphogenesis in the Basidiomycetes* (ed. M. N. Schwalb & P. G. Miles), pp. 135–165. Academic Press: New York.

Sen, S. (1992). Programmed cell death: concept, mechanism and control. *Biological Reviews* **67**, 287–319.

Smith, A. H. (1966). The hyphal structure of the basidiocarp. In *The Fungi*, **vol. II** (ed. G. C. Ainsworth & A. S. Sussman), pp. 151–177. Academic Press: New York.

Stephenson, N. A. & Gooday, G. W. (1984). Nuclear numbers in the stipe cells of *Coprinus cinereus*. *Transactions of the British Mycological Society* **82**, 531–534.

Takemaru, E. M. & Kamada, T. (1972). Basidiocarp development in *Coprinus macrorhizus*. I. Induction of developmental variations. *Botanical Magazine (Tokyo)* **85**, 51–57.

Waters, H., Butler, R. D. & Moore, D. (1975). Structure of aerial and submerged sclerotia of *Coprinus lagopus*. *New Phytologist* **74**, 199–205.

Waters, H., Moore, D. & Butler, R. D. (1975). Morphogenesis of aerial sclerotia of *Coprinus lagopus*. *New Phytologist* **74**, 207–213.

Watling, R. (1971). Polymorphism in *Psilocybe merdaria*. *New Phytologist* **70**, 307–326.

Watling, R. (1985). Developmental characters of agarics. In *Developmental Biology of Higher Fungi* (ed. D. Moore, L. A. Casselton, D. A. Wood & J. C. Frankland), pp. 281–310. Cambridge University Press: Cambridge, U.K.

Watling, R. & Moore, D. (1994). Moulding moulds into mushrooms: shape and form in the higher fungi. In *Shape and Form in Plants and Fungi* (ed. D. S. Ingram & A. Hudson), pp. 270–290. Academic Press: London.

Watling, R. & Nicoll, H. (1980). Sphaerocysts in *Lactarius rufus*. *Transactions of the British Mycological Society* **75**, 331–333.

Williams, M. A. J. (1986). *Studies on the Structure and Development of Flammulina velutipes (Curtis: Fries) Singer*. Ph.D. Thesis, University of Bristol.

Williams, M. A. J., Beckett, A. & Read, N. D. (1985). Ultrastructural aspects of fruit body differentiation in *Flammulina velutipes*. In *Developmental Biology of Higher Fungi* (ed. D. Moore, L. A. Casselton, D. A. Wood & J. C. Frankland), pp. 429–450. Cambridge University Press: Cambridge, U.K.

Wong, W. M. & Gruen, H. E. (1977). Changes in cell size and nuclear number during elongation of *Flammulina velutipes* fruitbodies. *Mycologia* **69**, 899–913.

Chapter 2

A new model for hyphal tip extension and its application to differential fungal morphogenesis

BYRON F. JOHNSON, GODE B. CALLEJA AND
BONG YUL YOO

Summary

Except for spherical (intussusceptional) wall growth during fungal differentiation, extension at the hyphal tip is the hallmark of the fungi. Even the tips of those highly modified fungi, the yeasts, are their primary sites of growth. The mode of tip growth of hyphae has been discussed casually in terms of softening and rehardening of the tip for about a century. As the branched nature of the structural polysaccharides of fungal cell walls became evident, a new paradigm for their molecular architecture became necessary. This was first explicitly incorporated into the endolytic model for tip growth. This model followed upon observations of lysis of fission yeast cells at their sites of growth after exposure to 2-deoxy-D-glucose. The next truly original model for tip growth was the 'steady-state' model, which attempted to explain wall rigidification by cross-linking between constituent polymers. These models have not been critically tested until recently. When faced with a critical test, some aspects of both models could be verified and some aspects of both models could be falsified. A new hybrid model is proposed here which stresses the need for controlled endolytic activity and resynthesis of polysaccharide at the tip (from the old endolytic model) and the need for the post-hardening and cross-linking activities from the steady state model. The new hybrid model utilizes the verifiable strengths of both models and avoids the falsifiable weaknesses of both.

Historical introduction

Except for some outstanding examples of intussusceptional wall growth in both the early literature (Errera, 1884, cited in Castle, 1953; reviewed by Castle, 1953; Park & Robinson, 1966; Bergman *et al.*, 1969) and more recently (see Chapter 1) occurring during fungal differentiation, extension of the cell wall at the hyphal tip is the outstanding characteristic of the fungi (Robertson, 1965a, b; Smith & Berry, 1978; Heath, 1990). Even in those highly modified fungi, the yeasts, the tips are the primary sites of growth. Some examples of this are *Saccharomyces cerevisiae* (reviewed by Robinow & Johnson, 1991), *Schizosaccharomyces pombe* (reviewed by Johnson, Miyata & Miyata, 1989), *Pichia farinosa* (Johnson & Gibson, 1966a), *Candida albicans* (Staebell & Soll, 1985), although *Mucor rouxii* (Bartnicki-Garcia & Lippman, 1969) is atypical and seems for now to be exceptional.

The mode of tip growth of hyphae has been discussed casually in terms of softening and rehardening of the tip for about a century (Marshall-Ward, 1884, cited in Gooday, 1977; Reinhardt, 1892). Most of the older discussions were presented either without knowledge of the molecular architecture of the fungal cell wall (e.g. the 19th century authors), or with apparent acceptance of the plant paradigm or concept for molecular architecture (e.g. Robertson, 1965a; see below, and Green's (1969) summary of plant structure). At any rate, there was no overt attempt by Robertson to distinguish between that prevalent paradigm and the notions of molecular architecture of the fungal wall that were just becoming prominent. Neither was Robertson (1965a) very explicit about lytic enzymes. Indeed, Trinci & Cutter (1986) suggested, 'The involvement of lytic enzymes in tip growth is not implicit in Robertson's model.' However, Robertson (1965b, 1968) alluded to the work of Nickerson's group on enzymes that 'soften' the tip as 'one of the most stimulating ideas in fungal morphogenesis' and whether he did or did not have in mind proper lytic 'softening enzymes' seems irrelevant in retrospect; we are satisfied to accept an implicit intent. However, we reemphasize that the context of the old plant concept of wall structure as it existed then implied that the softening enzymes had matrical molecules as substrates, not structural molecules. Hence Robertson's model stands separate from, and not as a precursor to a contemporary endolytic model for hyphal tip growth (Johnson, 1968a; see below).

As the branched nature of the structural polysaccharides of fungal cell walls became evident, a new understanding for their molecular architecture became necessary. That newer paradigm was first explicitly incorporated into a model for tip growth of fungi, the endolytic model, by Johnson (1968a). This model followed expressly upon observations by Megnet (1965) of fission yeast (*S. pombe*) cells lysing at their sites of growth after exposure to 2-deoxy-D-glucose (2DG), and was based upon experimental expansion and interpretation of those studies. The endolytic model has been adopted with some elaboration in detail by others (see below) but the next truly original model for tip growth came from Wessels & Sietsma

(1981). Wessels' 'steady-state' model also incorporated the modern paradigm for molecular architecture of the cell wall and attempted to explain its genesis by means other than endolytic activity.

These two models have co-existed since 1981, with no obvious basis for selecting between them. An appeal to Occam's Razor is fruitless: faced with the choice between the more complex endolytic model and the simpler steady-state model, the latter seems the obvious choice. But if the endolytic model must sometimes be invoked, as it must to initiate extension, then perhaps it alone is the more logical choice, as one mechanism must be simpler than two. Neither logic is even persuasive, let alone convincing. What was needed was a critical test. However, no critical test of either has appeared until recently (Johnson, Yoo & Calleja, 1995). Interestingly, when faced with a critical test, some aspects of both models could be verified and some aspects of both models could be falsified (Popper, 1983). Hence both models appear to have been assemblages of correct and incorrect notions.

Analysis of models

The endolytic model

The new fungal wall paradigm from the mid-sixties had structural molecules embedded in a matrix; it was a composite, as in the frequently drawn comparison with reinforced concrete. By contrast with the concept for plant wall (Green, 1969) in which the straight, unbranched cellulose-like structural molecules were enabled (in a relevant model) to 'slip' within an enzymically-softened matrix, enabling the stretching of the growing tip, the newer paradigm for hyphae had branched and cross-linked structural molecules that seemed neither able to slip upon one another nor able to slip within a 'softened matrix'. But the wall as a whole could stretch like a non-woven fabric, implicit earlier, explicit here. This new exemplar not only forbade application of the plant model, it demanded a new model that would enable extensile growth in the presence of branched and cross-linked structural molecules. According to that new model (Johnson, 1968a), extension occurs via semi-simultaneous hydrolysis of structural glucan molecules mediated by endoglucanase and by resynthesis of the glucan molecules to their former strength. Resynthesis could occur either by insertion of glucose or oligoglucan into the breaks (Johnson, 1968a) or by sufficient endwise synthesis in both directions from the breaks (Johnson *et al.*, 1995). The resultant endolytic model (Johnson, 1968a) attempted to rationalize the new paradigm, the close parallel of 2DG-induced lysis with

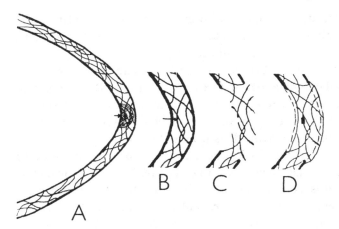

Fig. 1. Diagram of the endolytic model for hyphal tip extension. The only molecules depicted are cross-linked and branched structural glucan molecules. A, during extension, the site of exocytosis of a glucanase-bearing vesicle is noted at arrow, and the diffusion gradient of secreted lytic enzyme is indicated by concentric cross-hatching. The dense line at the inner surface of the wall suggests the periplasm; the dense line at the outer surface suggests the surface of the wall. Both entities might differ from species to species. B, magnified view at A, with the site of lytic attack indicated (arrow). C, stretched view, showing previously intact structural glucan molecule stretched apart at hydrolysed break, and the now-weakened non-woven fabric also stretched. Dense lines from A and B removed for clarity. D, integrity of the broken molecule restored by covalent insertion of oligoglucan into break (shown as heavy). Old wall dimensions shown by dashed lines, new wall dimensions shown by solid line.

the kinetics of extension (Johnson & Rupert, 1967), and the association of sites of 2DG-induced lysis of yeast cells with their apical sites of glucan synthesis during extensile growth (Megnet, 1965; Johnson, 1967, 1968a, b), and at the septum during cell division (Megnet, 1965; Johnson, Lu & Brandwein, 1974). Further evidence to support the model has been reviewed by Johnson, Calleja & Yoo (1977); Calleja, Johnson & Yoo (1981); and Johnson *et al.* (1989).

The effect of stretchability on tip growth is shown (Fig. 1). Thus, as endoglucanase hydrolyses one structural glucan molecule (Fig. 1B), the strength of the non-woven fabric approximated by the hyphal wall is diminished, and the ends of the cleaved glucan molecule are pulled apart (Fig. 1C) as the fabric stretches. Resynthesis either by insertion (see above; illustrated, Fig. 1D) of oligoglucan or by synthetic extension of the divided molecule (not shown) generates a larger fabric, having the full strength as before, but now capable of greater stretch. That is, the tip has grown. We assume that the ends of cut structural glucan molecules will be pulled apart in proportion to their distance from the periplasm, thus a cleaved molecule at the surface of the tip will require a longer oligoglucan insert to restore intactness. Structural studies (Aronson & Preston, 1960; Hunsley & Burnett, 1970; reviewed by Aronson, 1981) showed glucan molecules persisting even on growing hyphal tips. The endolytic model requires a relevant endoglucanolytic activity. Such an activity was later found by Barras (1969, 1972) and confirmed by Fleet & Phaff (1974).

The merits of the endolytic model are (i) that it takes into account the new concept of molecular architecture; (ii) that it readily allows rationalization of a variety of lytic phenomena at the hyphal tips, including the 2DG results of Megnet (1965) and others (Johnson & Rupert, 1967; Johnson, 1968a, b; Moore, 1969; Biely, Kovarik & Bauer, 1973; Johnson *et al.*, 1974; reviewed by Moore, 1981), the lytic and enzymological studies of Bartnicki-Garcia & Lippman (1972), Bartnicki-Garcia (1973), Gooday (1977, 1983), Fèvre (1979), Rosenberger (1979), (iii) and it does these in a simple manner. The endolytic model has deficiencies as well. For example, (i) it treats the wall as though it were a homogeneous entity, i.e. as though the wall were isotonic (equally stressed) everywhere except at the growing region; (ii) it is based only upon indirect experimental data, lacking persuasive confirmatory experiments; and (iii) its simplistic nature leaves out many relevant aspects of hyphal tip extension, such as secretory vesicles, the cross-linking so important to Wessels' concepts, and roles for the cytoskeleton, especially actin. Yet cross-linking at the tip poses no problem to the endolytic model.

Biological systems are usually well provided with repair systems (discussed by Heilbrunn, 1952). A molecular example is shown in Table 1 in McMacken, Silver & Georgopoulos (1987): DNA polymerase I, a DNA repair enzyme, is expressed at *ca* 300 molecules per *E. coli* cell and DNA polymerase III, the DNA replicative polymerase, at *ca* 20 molecules per cell. This ratio of 15:1 seems of the right order for expression of a repair function over expression of an ordinary cell-cycle function. Thus, we make the explicit point that endolytically-cleaved structural wall molecules would seem like a wound to the hyphal cell, and would require repair by repair systems which are in excess. Although repair must be kinetically exuberant, it too must respond quantitatively to regulation so that one endolytic cleavage does not stimulate massive wall synthesis. Co-regulated generation of substrate (cleaved ends) and product (wall integrity restored) implies 'balance', but not 'delicate'. Notions of a biochemical 'delicate balance' (Bartnicki-Garcia & Lippman, 1972) of lytic and synthetic enzymes appear to us to be very unbiological (*vide* the ratio between DNA repair and synthesis quoted above). There can be no objection to a sense of equilibrium, but stoichiometry seems improbable. Unless stoichiometric balance of expression and/or potential activity of the relevant enzymes can be unequivocally demonstrated we consider these 'delicate balance improvements' of the model to be of negative value. One purported 'confirmation' of the delicate balance (Kritzman, Chet & Henis, 1978) actually shows a balance between mostly exoglucanolytic activity and synthesis because the assay was not specific for endoglucanolytic activity, most of the measured glucose being released by *exoglucanolytic* activity. Barras (1972) has shown that purified endo-β-glucanase releases reducing sugar but not glucose. Although Johnson *et al.* (1977) suggested roles for exoglucanolytic activity at cell division and at conjugation (see below), no one to our knowledge has proposed a role for exoglucanolytic activity in extension at the tip.

The steady-state model

It is to Wessels and his group that we owe a debt for many of the data that led to a new (steady-state) model (Wessels & Sietsma, 1981) and its adaptations (reviewed by Wessels, 1993) as new data accumulated. According to the steady state model, extension occurs when the pre-softened wall at the hyphal tip is stretched thin along its axis; the wall is rethickened, restored, by vesicular exocytosis of proteins and polysaccharides as wall precursors, also at the tip. The hardness (strength) of the wall is re-established via a cross-linking mechanism that is initiated minimally at or near the tip and continues progressively further back in the hyphal wall (see Fig. 2). Pre-softening is catalyzed, of course, *via* an endolytic mechanism that according to the model is briefly called into action solely to initiate the growth mode but not to sustain it.

The merits of the steady-state model include (i) that it takes into account the new concept of molecular architecture; (ii) that it readily allows rationalization of a variety of synthetic and exocytotic phenomena occurring at the hyphal tip; and (iii) that it involves a cross-linking mechanism for rigidifying and maturing the wall away from the extending tip; the wall in this model is obviously not homogeneous. Among its deficiencies are (i) that the kinetics of 2DG-induced lysis (assuming that 2DG-induced lysis quantitatively reflects endoglucanolytic activity; see above) and 2-deoxy-2-fluoro-D-glucose-induced lysis (Biely, Kovarik & Bauer, 1973) seem almost inexplicable except at initiation of extension; and (ii) that it seems to depend

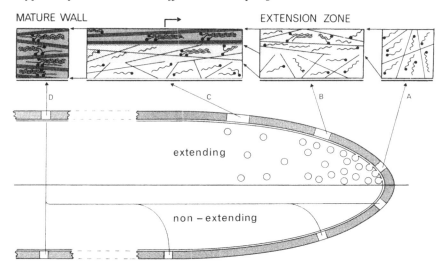

Fig. 2. The steady-state model of apical wall growth. For the extending apex (vesicles fusing with the plasma membrane) the schematic drawing shows the stretching of the wall and the addition of new wall material from the cytoplasmic side, maximally at the extreme tip. Newly added wall consists of chitin chains (straight lines) and (1- > 3)-β-glucan chains (wavy lines). Other wall components are omitted because they are less important as structural elements. With time, these two polymers interact to form chitin microfibrils and β-glucan triple helices while covalent linkages occur between chitin and β-glucan chains which also become branched. At the very tip (a) the wall is minimally cross-linked and supposed to be most plastic. Subapically (b) a wall volume added at the very tip is stretched and partially crosslinked (increased shading) while new wall material is being added from the inside to maintain wall thickness. Note that wall material moves from the inside to the outside and that the wall at the outside is always the oldest. Areas with the same shading in (a), (b), and (c) represent equal volumes of wall but note that in reality there are no discontinuities in the wall. At (c) the average crosslinking has proceeded to the point that the wall hardly yields to turgor pressure but it has not yet maximal strength. Crosslinking continues in the tubular part of the wall till completion far behind the apex (d). In a non-extending apex (no vesicles present), the steady state breaks down because the steady-state amount of plastic wall material is no longer pushed out of the zone of rigidification. Crosslinking between the wall polymers now occurs over the whole apex (d). From Wessels (1993), reproduced with permission.

upon a biophysical balance of flowing wall precursors before their rigidification if the tip is not cross-linked (as is implied by the discussion in Wessels (1993) and Fig. 2) which is just as difficult to visualise as is the biochemical 'delicate balance' discussed above.

The actin knob model

A recent cytological paper (Mulholland *et al.*, 1994) very briefly discusses application of deductions from their interesting but enigmatic micrographs to a model for extension. The brevity and especially the lack of generality of the model preclude further discussion except later under actin (see below).

Testing the endolytic and the steady-state models

Over the years since these two models were first proposed, a variety of results have been amassed that purport to support one or the other of the two models. Until recently, no critical experiment has been reported that could falsify (Popper, 1983) either model.

In fact, it has not been easy to devise falsifying experiments. Consider secretion vesicles, for instance. Cortat, Matile & Wiemken (1972) showed in *Sacch. cerevisiae* that glucanases were secreted from vesicles at the tips of the buds. Miyata & Miyata (1978) showed the close association of cellular extension and secretion in the fission yeast, *S. pombe*, and these results were confirmed and extended to *Sacch. cerevisiae* by Field & Schekman (1980). Also, Novick, Field & Schekman (1980) devised a selection scheme for the isolation of *sec* (secretion) mutants whose study allowed elaboration of the secretion pathway in eukaryotes. These secretion vesicles were intrinsic to the thinking behind the Wessels' models and were the essence of a model by Prosser & Trinci (1979), but were rarely discussed in connection with the endolytic model (some exceptions were Cortat *et al.*, 1972, and Johnson *et al.*, 1977). Nevertheless, there is no suggestion that the periplasm becomes loaded *de novo* with wall precursors and lytic enzymes, hence a comparison of rates of fusion of secretion vesicles to the plasmalemma with the rates of extension can falsify neither.

For other examples, we note that the cross-linking aspect of the steady-state model is almost overwhelmingly convincing, but the range of substrates and enzymes and co-factors involved must be so great that any specific assay might be entirely misleading (unless one could know with confidence which activity was the crucial one!). The same is true for testing the strength-restoration repair step of the endolytic model; it is entirely possible that the crucial enzyme has not yet been characterized or isolated.

On the other hand, the endo-β-glucan hydrolase of Barras (1969, 1972) is a test candidate. According to the steady-state model, activity should be high at the moment of initiation of extensile growth, but should shortly fade to zero. Once the tip is pre-softened, continuous endolytic activity is superfluous according to Wessels (1993). By contrast, continuous endo-β-glucanolytic activity is required by the endolytic model (Johnson, 1968a); the greater the rate of extension, the greater the endolytic activity. One might select between the models if one could do the proper assay, quantitatively on single cells, but that seems not to be possible yet. Barras (1972) did show that activity paralleled more-or-less the growth curve of bulk cultures of fission yeast cells, but while that is convincing enough from the point of view of the endolytic model, it is also persuasive from the steady-state point of view. The faster the fungal colony expands, the more branching will occur, and branching requires initiation via endolytic activity. These points were brought out explicitly by Thomas & Mullins (1967) and recently expanded by Money & Harold (1992, 1993). Thus the Barras (1972) experiment is useful in that it suggests the relevance of the particular enzyme, but it was not a critical experiment.

Because it seemed impossible to devise a critical experiment at the chemical/biochemical level at which the models were posed, we wondered if the models could be tested at a less sophisticated level (Johnson *et al.,* 1995). Our approach was based on unpublished observations of fission yeast walls that had been smashed between glass beads or between the beads and the vessel wall. New quantitative ballistic experiments were performed and the results fell into two categories: those that seemed germane to the specific growth habit of *S. pombe* and results that seemed to be of general interest, germane to the models or to predictions from the models.

Briefly, the fission yeast cell grows by pseudo-exponential extension (Miyata, Miyata & Johnson, 1988) at one (Mitchison, 1957; Johnson, 1965) or sometimes both ends (Johnson, 1965; Mitchison & Nurse, 1985; May & Mitchison, 1986; H. Miyata *et al.,* 1986, 1988, 1990). A consequence of the pseudo-exponential nature of extension is that, by and large, the rate of extension of this cell is a function of the length of the cell (Johnson, 1965; 1968c). Also, growth in bulk culture is moderately unbalanced (Campbell, 1957) so that the average length of cells declines through the log phase, after an exponential extension in the lag phase (Mitchison, 1957; Johnson, 1968c; James *et al.,* 1975). In our smashing experiments the frequency of ruptures at the extensile end was found to increase through the lag phase and to decline through the log phase (Johnson *et al.,* 1995). That is, the fragility was correlated with the mean rate of extension in both circumstances. In addition, more recent studies (Sabrina Piombo, personal communication) show that the frequency of rupture at the extensile end is correlated with its rate of extension during the cell cycle. These results might have been predicted by both the endolytic and the steady-state models.

What is of more general interest is the nature of the ruptures and their relevance to the models. A high proportion of the ruptures were seen by light and electron microscopy as out-curled. Out-curl was characteristic of ruptures at both extensile regions and non-extensile regions (mostly but not exclusively cylindrical wall; Johnson *et al.,* 1995). A prediction based on Wessels' steady-state model (1993) is that cylindrical wall could have a gradient of tension, increasing from inside to outside, established by the operation of the cross-linking system: the older the wall domain, the more it cross-linked. Thus, out-curl should be apparent at those ruptures (and is: Johnson *et al.,* 1995). An interesting aspect of the steady-state model was thereby confirmed. The endolytic model does not appear to generate a prediction for the nature of ruptures of cylindrical wall, hence must be considered here to have been found inadequate by that test.

On the other hand, out-curl at the extensile end seems inexplicable by the steady-state model. The gradient of tension in cylindrical wall is primarily a consequence of cross-linking. But the tip as conceived by Wessels cannot simultaneously be markedly cross-linked and plastic enough to expand. As stated by Wessels (1993, page 404): '... a plastic mixture of wall polymers [extruded] maximally at the extreme tip ... allows the wall to expand maximally at this site and to grow away from the zone of rigidification...' Inexplicable means falsified, or again, inadequate by that test.

With regard to the endolytic model, Johnson *et al.* (1995) proposed that diffusion of the secreted autolytic enzyme molecules would generate a concentration gradient of endolytic activity in the wall at the tip. Furthermore, the steepness of that gradient would be exaggerated by the natural binding of enzyme to substrate, so that endoglucanolytic activity would be highest near the plasmalemma and lowest at the exterior surface of the tip. Thus, inherent to the endolytic model, one might expect a gradient of tension at the extensile tip established by the endolytic enzyme, essentially mimicking the gradient of tension in the cylindrical wall established by the cross-linking known to occur there. Accordingly, the arguments about the gradient of tension in the cylindrical wall that falsified the endolytic model there now confirm it at the tip.

Another factor to consider is that the relative frequency of out-curling at the tip declines as a function of time after inoculation (Johnson *et al.,* 1995). The decline is correlated with a decline in mean cell length through the exponential phase of growth, hence with a decline in the mean rate of extension at the tip (see above). It does seem like a logical prediction from the steady-state model that the slower the extension rate, the greater would be the cross-linking leading to tension at the surface of the wall, hence the greater the tendency for ruptured walls to out-curl hence the greater the out-curl. Accordingly, by this test also, the steady-state model fails, its prediction being contrary to the trend of the data. Coincidentally, data that lead to the conclusion that endoglucanolytic activity persists through extension also falsify that element of the steady-state model that holds that endolytic activity is deemed unnecessary during continuous extension. By contrast, the steepness of the concentration gradient of lytic enzyme (endolytic model) at the tip would be less pronounced as the tip growth slows. Hence a decreased tendency for tip ruptures to out-curl late in the log phase would be predicted, quite in accord with the data.

Thus, both models were simultaneously confirmed and falsified by the same critical experiment (Johnson *et al.*, 1995). Because the confirmations of the two models and the falsifications of the two models were both complementary, we suggest that the surviving portions of both models be fused into a new model: a hybrid model which is presented below.

A new hybrid model for extension

From the existing steady-state model (Wessels, 1993) we retain the concepts of exocytosis of relevant enzymes and their substrates and preformed carbohydrates from secretory vesicles at the tip, the verified out-to-in gradient of cross-linking and hardening among the structural (carbohydrate) molecules, and process regulation involving stretch receptors. Specifically rejected among the steady-state model concepts is that of merely transient endoglucanase activity to establish a pre-softened tip. Among the concepts retained from the endolytic model (Johnson 1968a, b; Johnson & Rupert, 1967; Johnson *et al.*, 1977; Johnson *et al.*, 1995) is continuous endoglucanolytic activity expressed in proportion with the rate of tip extension. Also retained is resynthesis of endolysed structural molecules, either by insertion of glucose or oligoglucan across the break (Johnson, 1968a), or by linear synthetic addition to the new, endoglucanase-generated ends of the glucan molecules (Johnson *et al.*, 1995), and the capacity to stretch which we now see as being very important. Turgor can easily stretch the surface area of a wall to twice its equilibrium relaxed area (Johnson *et al.*, 1995). Obviously a model that includes a constant capacity for endolytic cleavage can accommodate the presence of structural fibrils across the surface of the tip (Aronson & Preston, 1960; Hunsley & Burnett, 1970; Aronson, 1981). One suspects that stretch receptors (Wessels, 1993) regulate resynthetic activity to control the extent of resynthesis; the rate of resynthesis will reflect the hyperabundance of resynthetic repair enzymes as discussed above. Also retained is the concentration gradient of endoglucanolytic activity generated by diffusion from the point of exocytosis and amplified by binding of enzymes to their immobilized substrates within the wall at the tip (Johnson *et al.*, 1995, and above). The new hybrid model is diagrammed in the discussion of conjugation, below.

Adjuncts to the model

The role of turgor

Concepts of the role of turgor in extension run the gamut from it being absolutely essential (Ray, Green & Cleland, 1972) to it being trivial (*Achlya bisexualis* can grow in the absence of turgor (Money & Harold, 1992, 1993)). In the usual situation, turgor is present and should be expected to participate in extension as described by Ray *et al.* (1972) and by Wessels (1993). Where specific examples appear to be intrinsically different, as in *Saprolegnia ferax* (Kaminskyj, Garrill & Heath, 1992) and *Achlya* (Money & Harold, 1992, 1993), they probably represent specific mechanisms for specific circumstances rather than generally applicable patterns.

The role of actin

Biochemical studies of actin go back over half a century (Engelhardt, 1942) but cytological studies of actin in fungi are so new that notions of actin function could change every year or so. Current emphasis (proposal) on structural reinforcement of the hyphal tip (e.g. Jackson & Heath, 1990) is based on too narrow a focus, the tip only, and on anxiety that the extending tip might not be stróng enough to resist turgor (Mulholland *et al.*, 1994). We appreciate that it is only natural for the tip to attract the attention of mycologists ('The key to the fungal hypha lies in the apex' Robertson, 1965*a*), but yeasts are fungi and their division highlights something different: actin plaques at the septum as well as the tip. Their presence in the fission yeast, *S. pombe,* has been demonstrated using light microscopy (Marks & Hyams, 1985; Marks, Hagan & Hyams, 1987; reviewed in Robinow & Hyams, 1989) and in the budding yeast *Sacch. cerevisiae* using electron microscopy (Mulholland *et al.*, 1994). What can be the function of these septal actin plaques? Structural reinforcement does not seem so imperative here; not only is the septum itself fairly tough (often double the ordinary wall thickness), it should not ordinarily be facing the problem of resisting turgor, simply because the turgor pressure must be about equal on both its sides. Another interpretation which seems likely comes from the recognition that the tip and the septa are very busy places during their biogenesis. Thousands of vesicles reach those sites for terminal exocytosis. The cytoskeleton will surely have organizational responsibilities for this transport, and actin will be involved. The detailed immunocytochemical analysis of plaques in *Sacch. cerevisiae* (Mulholland *et al.*, 1994) makes it evident that actin 'cables' can have firm attachment at both the tip and the septal regions of wall synthesis. Thus, actin might be functioning especially in transport.

Furthermore, the behaviour of actin plaques in the fission yeast (as reviewed in Robinow & Hyams, 1989) is reminiscent of Mitchison's statement (1957) that '... the point of deposition of the new cell wall in each daughter cell [just after division] moves back from the new end formed from the cell plate to the old end of the cell,' something we have referred to as Mitchison's Rule (Calleja, Yoo & Johnson, 1977a). Also worth noting in the context of comparison of septal and apical wall growth is that not only is there no pressure difference to deal with, but the mechanism of septum formation is totally different from the mechanism of tip extension (Johnson *et al.*, 1977). Accordingly, we conclude that the cytochemical observations suggest that actin at the hyphal tip probably functions very much as it does at the septum: in vesicle traffic control and biosynthesis, with no strengthening function at all. A recent review by Bretscher *et al.* (1994) emphasizes these points.

Function of the new hybrid model

Monaxial, in-line extension

Positive, monaxial extension exists in two forms: hyphal and continuous, or yeast-like and determinate. Application of the new hybrid model to continuous, positive, monaxial extension seems evident without further discussion, because that is the form of extension for which the new model and some older ones were devised.

A small number of examples suffices to illustrate what is commonly regarded as determinate, positive, monaxial extension. First, initiation of budding in the first adult cell cycle of diploid *Sacch. cerevisiae*. And second, resumption of extensile growth at a previously extensile end by newly divided fission yeast (*S. pombe*) cells. Old walls of both these yeasts grow non-accretionally during every cell cycle (Beran, Streiblová & Lieblová, 1969; Johnson & Lu, 1975), but the resultant increases of volume are not

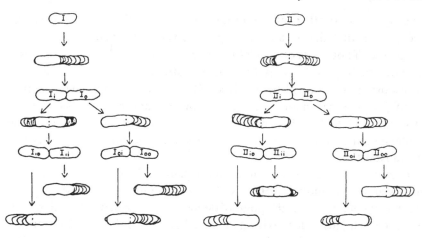

Fig. 3. Growth of cells which were followed by superimposing time-lapse photomicrographs every 20 min. I-cell and II-cell lines, initial cells grew at one end or at both ends, respectively. o, Outside cell; i, inside cell. Cells I_ρ, $I_{\rho\rho}$, $I_{\kappa\rho}$, $I_{\kappa\kappa}$, II_κ, II_ρ, $II_{\rho\kappa}$, $II_{\rho\rho}$, and $II_{\kappa\rho}$ all follow Mitchison's rule. Cell $I_{\rho\kappa}$ violates Mitchison's rule. Cells I_κ, II, and $II_{\kappa\kappa}$ are all bipolar. By convention, cell I is stated to be monopolar. Revised after H. Miyata *et al.* (1986), with permission.

large, less than 25% per cell cycle, hence we can consider the growth of both cells to be determinate, in accordance with custom.

In the diploid *Sacch. cerevisiae* example, a new adult initiates its first bud at the end that was extensile during its own budding growth. Hence the new growth is positive, and is usually monaxial, i.e. extension on precisely the same cellular axis. This pattern has been known since the time of Winge (1935), was expanded by Townsend & Lindegren (1954) and genetically rationalized by Chant & Herskowitz (1991, but with a different convention for axiallity). The new budding cycle begins only after a period in which the new adult has grown post-scissionally (Johnson & Gibson, 1966b; Hartwell & Unger, 1977) to attain the 'critical cell size' (Johnston, Pringle & Hartwell, 1977). This growth of the new adult during which its progenitor was initiating another new budding cycle is typical yeast unbalanced growth (Campbell, 1957; Johnson & Gibson 1966b; Flegel, 1978; reviewed by Robinow & Johnson, 1991).

The fission yeast example was made famous by Mitchison's classic paper (1957), and reexamined in detail (Fig. 3) by H. Miyata *et al.* (1986). Briefly, this sausage-shaped yeast extends at one or both ends and then divides. Division generates two sibs, each with a new (division-generated, scarred) end and an older end. Mitchison's Rule (Calleja *et al.*, 1977a) says that the older of the two ends most often initiates extension first, but sometimes both ends extend simultaneously, and some cells violate Mitchison's Rule (H. Miyata *et al.,* 1986). Ignoring these less common exceptions, and restricting discussion to older ends that were themselves extensile in the previous cell cycle, we find the example completely in harmony with the diploid *Sacch. cerevisiae* example immediately above.

These two examples have much in common with continuous hyphal extension: they are essentially continuous but with interruptions and may be better regarded as being 'discontinuous' rather than determinate. In both of these examples, it should be emphasized that these cells reinitiate monaxial growth precisely where the cell extended in a previous cycle. These cells would seem to have a 'memory' and we suggest that Marks *et al.* (1987) have shown a different example but a satisfactory suggestion for a virtual memory. Thus, they showed that some fission yeast cells had an actin ring, presumably apposed to the plasmalemma, that marked the last pre-mitotic position of the dividing nucleus. Functioning as a memory, it indicated to the cell the site for the new septum.

We propose simply that when the actin plaques 'migrate' from the extensile end (Marks & Hyams, 1985; Mulholland *et al.,* 1994) to the site of the new septum for divisional purposes, some crucial 'memory' vestiges of actin remain to mark the site for the next cycle of extension. If not actin, then some actin-associated structure might have been deposited and remained there. In much the same manner, the new end formed by the divided septum (*S. pombe*) is memory-marked so that when it initiates extension, that extension is remarkably monaxial. Rare exceptions to monaxiallity will be discussed below.

In essence, the putative memory-marker at the tip should serve as an assembly region for the actin plaques or networks (Jackson & Heath, 1990) that are sure to function (in transport) during the next extension cycle. Presumably by having the vesicular components (Prosser & Trinci, 1979) of cell extension delivered to the correct place at the correct time, the endolytic aspects of extension, as depicted in the new hybrid model, are activated at precisely the correct monaxial site.

Extension at septa

Strictly, septation, or cell division, is not germane to this review on extension. However, septa might persist, and they can become sites for initiation of branching, itself the *de novo* initiation of extension: '... [I]n moulds which do form septa, branching is usually associated with their presence; depending on the species...' (Trinci, 1979). On the other hand, scission might occur, generating a scission scar, and that might be either a forbidden site for extension (e.g. birth and bud scars of budding yeasts) or a favoured site for extension (e.g. the necks of apiculate yeasts, or terminal scars on fission yeasts, or the second or higher cycle sporangium formation by *Saprolegnia* [see Fig. 49 in Alexopoulos, 1962]). Finally, there are hyphae that branch but never have septa: 'Of course, branching in moulds is not invariably associated with septation.' (Trinci, 1978).

Favoured sites for extension are no problem; the putative memory-marker from septation enables operation of precisely the same mechanism as described above for monaxial extension.

Forbidden sites seem inexplicable now. It is worth noting that old fission scars of *S. pombe* were held to be forbidden, or at least anomalous, sites for future septa (Johnson *et al.*, 1982) until M. Miyata, Miyata & Johnson (1986b) showed that notion was incorrect. However, in ordinary budding yeasts, birth and bud scars remain as tokens of forbidden sites for extension. But if one's first thought is that a favoured site retains its virtual memory and a forbidden site has obliterated its virtual memory, one should recall the 'string-of-beads' appearance of successive scars in Streiblov's famous picture (1970). This clustering of successive budding sites must indicate the functioning of a residual virtual memory.

Forward branching at old septa is not so different. The memory-marker from septum formation putatively functions to identify the general site for extension, but the septum itself seems forbidden, possibly because of the turgor of the apposed viable protoplast, hence the extension mechanism is angularly diverted. At first sight, backward, or negative, branching at septa might seem to be fundamentally different. But the backward hyphal branching (Fig. 4) described by Haskins (1967) and the backward branching forms of hyphal *S. pombe* (Fig. 5) genetically described by Sipiczki, Grallert & Miklos (1993), and also phenotypically described by Johnson & McDonald (1983), do not seem so different if considered in terms of initiating the extension mechanism at old septal actin plaques as discussed above.

Branching remote from septa will obviously utilize the extension mechanism, but the basis for selecting the site of application remains inexplicable. Perhaps the site of branching was once a site of mitosis that retained an actin memory. Just as likely is the possibility that site selection for all remote branching is somehow illustrated by the complex bud initiation schemes of *Sacch. cerevisiae* that have been so extensively studied (see Chant & Herskowitz, 1991; Klis, 1994; Cid *et al.*, 1995).

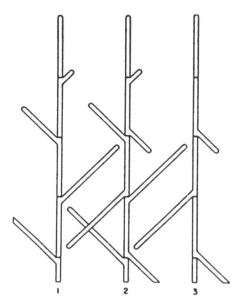

Fig. 4. Diagrams showing three patterns of hyphal branching in the fungus P.R.L. 2176. 1, forward branching pattern, with branches initiated 'immediately below the septum at the distal end of the cell.' 2, second branches also, initiated 'immediately in front of the septum.' 3, 'only backward branches develop, and always away from the proximal end of the cell just in front of the septum.' These examples of branching all occur at septa. From Haskins (1967), reproduced with permission.

Conjugation

Fundamentally different in its genetic control must be homothallic conjugation via dissolution of a preexisting septum between two recently divided sibs. For instance, *Schwanniomyces occidentalis* (Kreger-van Rij, 1977) and *Schizosaccharomyces octosporus* (Ashton & Moens, 1982) both have sib-sib trans-septum conjugation through cell ends that were generated after the mitosis during which one cell's mating type was switched. We are assuming that conjugation tube ('beak' in the terminology of Ashton & Moens, 1982) formation occurs *via* an extension mechanism.

Conjugation requires paired cells to make contact either stochastically or by outgrowths directed toward each other, e.g. shmoos in *Saccharomyces cerevisiae* (MacKay & Manney, 1974). Effective stochastic contact implies that any region of wall might become fusion organelle; in a sense, the cell is regionally promiscuous. Remarkably, regional promiscuity is repressed rapidly, because multiple conjugations per cell are infrequent (Calleja *et al.*,

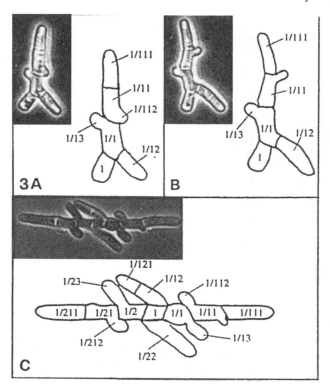

Fig. 5. Growth pattern in microhyphae of *sep*1-1. (A) and (C) cell lines that grew predominantly at their old ends. (B) a mixed line containing a cell (1/11) that violated Mitchison's rule, because it started growing at its new end. The tracings show the chronological order of cell divisions. Number 1 marks the mother cell of the hypha. The sibs are numbered in the order of their birth. For example: 1/1 and 1/2 are the first and second daughters of the initial cell, respectively; 1/11, 1/12 and 1/13 are the first, second and third sibs of the cell 1/1; etc. Bar = 5 μm. From Sipiczki *et al.* (1993), reproduced with permission.

1981). The paired cells must remain in close contact, and form covalent bonds as the first overall step in fusion of their walls (Calleja *et al.*, 1981). The new hybrid model can probably be construed to function in conjugation but will need supplementation. The early steps described above might vary from species to species; indeed, even the formation of covalent bonds between the paired cells might vary.

We presume that the relevant covalent bonds join wall polysaccharide molecules of the paired conjugants. Two obvious mechanisms are: (i) cross-linking between touching walls, and/or (ii) the formation of continuous glucan molecules where ends are in close enough proximity to allow the 'resynthesis' mechanism to operate (cross-resynthesis). An aspect of both mechanisms during extensile growth not discussed above is the need to have them restricted to functioning strictly within the confines of the cell wall. It would be disastrous to have the wall cross-link or otherwise covalently bond to any and every foreign polysaccharide molecule that happens to float close by. If either of the two mechanisms is to function in forming conjugal covalent bonding, that restriction must be partially relaxed as part of the conjugation process. Perhaps relaxation is induced while the erstwhile conjugants are held in contact. We cannot imagine how these controls function, either in restriction or its relaxation.

Because conjugants have been extending toward each other, their extension machineries are in full flight, which means that the normal wound-healing activities of resynthesis enzymes will pertain. Any two ends of wall polysaccharide molecules that are found near each other presumably will activate the insertion or glucan-extending mechanism; we suggest that in this special circumstance, the two molecular ends must originate from the walls of two different cells, so that molecular continuity between the two cells becomes established (cross-resynthesis, Fig. 6A). The first bond will act as a focal point for more resyntheses/cross-links to be established (Fig. 6A & B), in analogy with a two-dimensional zipper. Being plastic, the walls will readily flatten on each other (Fig. 6B & C) to fuse. Meanwhile, the extension mechanism as a whole is now functioning in a non-extensile circumstance in which it might well be regulated to reshape the fused walls. Obviously there are many structural molecules to cleave, many new bonds to form (Fig. 6B), much cross-linking to be established (Fig. 6C), and some wall materials to be deleted, probably by exoglucanolytic activity (Fig. 6D), before the membranes can be brought into contact.

Perforation will change things (Fig. 6E). Pre-perforation bonding between the conjugants must be complete and effective; it seems obvious that once the walls are perforated, forces (turgor, etc.) will stretch its inner surface. The new ends of every cleaved structural polysaccharide molecule will quickly be pulled so far apart that any resynthesis will more likely be end extension rather than oligosaccharide insertion. Once again, post-perforation morphogenesis will require much cross-linking, much vesicular addition of matrix-type materials, constant monitoring of stretch, and much endolytic activity. One can also see, almost from the beginning, a role for exoglucanolytic activity, to clip off ends of polysaccharide molecules, so that the remaining walls between the pre-conjugants can be removed

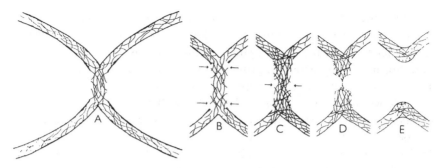

Fig. 6. The new hybrid model for hyphal tip extension applied to conjugation. Dense lines representing outer surface and periplasm are removed for clarity. A, the tips of the obverse cells are depicted as flattened together after endohydrolysis and cross-linking (see text). B, the area flattened together via 'cross-resynthesis' after endohydrolysis and by cross-linking is considerably enlarged. Paired arrows indicate where cross-linking must become even more extensive (see C) before perforation can safely begin. C, extensive cross-linking and structural glucan cross-resynthesis are now evident near the exterior surface of fused cells. Paired arrows indicate where perforative erosion will occur as the consequence of glucan endohydrolytic activity followed by glucan exohydrolytic activity. D, concentric erosion patterns (see Fig. 1A) about to meet. Glucan exohydrolytic activity after endohydrolysis obviously supersedes resynthesis in this region. E, Once the conjoined areas of wall are perforated, and the cytoplasts are fused, the remnants of wall remaining are quickly removed. Turgor will function here by activating stretch receptors in this narrow conjugal neck and will stimulate massive restructuring *via* endohydrolysis and resynthesis, *via* extensive cross-linking, and *via* exohydrolytic removal of the ends of molecules deemed to be superfluous.

completely. Simultaneously, the fusion has become so complete that there remains no evidence of earlier cellular identities at the site of junction.

We have proposed a similar but less detailed model for conjugation before (Johnson *et al.*, 1977). We have also dealt at some length with the propensity of conjugal morphogenetic systems for lytic self-destruction (Calleja *et al.*, 1977a & b, 1981; Johnson *et al.*, 1977).

Protoplasting revisited

Two aspects of the production of protoplasts are relevant here. First is the fact that one can release protoplasts using a cocktail of enzymes that, on the face of it, is inadequate: it will not completely dissolve the target wall. The classic example is the use of snail juice, mostly β-glucanase activity but no demonstrable α-glucanase activity, to release protoplasts from fission yeast

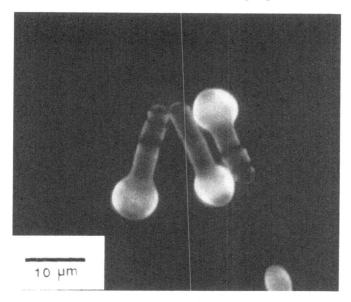

Fig. 7. Fluorescence photomicrograph of deformed fission yeast cells (*Schizosaccharomyces pombe*) stained with Calcofluor after incubation with aculeacin A. (Courtesy of Dr M. Miyata).

cells (*S. pombe*) that have α-linked glucan functioning to some extent as structural glucan as shown by the round-bottom flask cells of the Miyatas (see below, Fig. 7). It is difficult to imagine how all this can happen unless the target cell has α-glucanase as an endolytic enzyme, and unless the disorganization of the wall induced by the β-glucanase from the snail juice can lead to activation or release of the putative α-glucanase. The second aspect is the widely known fact (e.g. Russell, Garrison & Stewart, 1973) that fast growing cells are easily protoplasted and that cells growing slowly or not at all require more effective cocktails. Stationary phase cells are notoriously difficult to protoplast, sometimes being completely refractory. This aspect might be relevant due to completed cross-linking (Wessels-type) or for reasons related to the first aspect: disorganization of the wall induced by the cocktail activates/releases the endogenous endolytic enzymes which then participate in the dissolution or at least the weakening of the cell wall. In the context of the new hybrid model, the faster growing cells have more activated endogenous endolytic enzymes than do the slow growing or non-growing cells, thus are able to 'participate' more vigorously.

Perturbations

Saltatory extension rates

López-Franco, Bartnicki-Garcia & Bracker (1994) have quantitated 'pulsed growth' of hyphal tips of seven different species of fungi. They used modern video analytical tools capable of 'superresolution' (Inoué, 1989) but only in conjunction with phase-contrast optics whose resolution limitations have been evident for several decades, a failing emphasized once again by Salmon, Walker & Pryer (1989). Indeed, Inoué (1989) suggests, '... the conventional limit of resolution ...provides a reasonable criterion for defining the smallest distances resolvable in a complex structure.' The saltations in *Rhizoctonia solani* and *Fusarium culmorum* measured by López-Franco *et al.* (1994) were of low enough magnitude that all of the superresolution of their video analytical equipment might have been incapable of conviction. But saltatory extension of *Gilbertella persicaria*, the best among the seven fungi examined, could certainly have been resolved with conventional photomicrography. Indeed, the authors attribute similar findings to earlier authors, among them Trinci (1978) and Robinow's classic study of *Basidiobolus ranarum* (1963). In essence, those similar findings were mostly qualitative, a consequence of the different emphases imposed by their authors. Hence López-Franco *et al.* (1994) by their quantitative approach have generalized the saltatory circumstance; their emphasis upon this above all else is new, and draws attention to an important phenomenon. We conclude that the saltatory extension rates are real, but there may be new considerations. Their results showing variation of rates of extension of *G. persicaria* in a regular pattern (see Fig. 2 in López-Franco *et al.*, 1994) can be re-examined over three typical cycles of waning and waxing rates. The cyclic decreases of rate averaged 0.44, 0.32, and 0.33 μm sec^{-1}, the mean being 0.36 μm sec^{-1}. The cyclic increases of rate (matched with the above) averaged 0.20, 0.26, and 0.26 μm sec^{-1}, with a mean of 0.23 μm sec^{-1}. These rate oscillations may reflect periods wherein autolysis prevails over resynthesis (where the rate is increasing) followed by periods wherein resynthesis prevails over autolysis (where the rate is decreasing), but obviously, both carry on continuously and simultaneously during extension.

Finally, perhaps saltatory extension rates should not be surprising at all. We suggest that it might be more surprising if so complex a process as extension were to run perfectly smoothly, always.

Malforming morphogenesis

In recent studies, it has been noted that fungal growth may become disorganized, and that unusually weak and malformed hyphal walls, some actually even shaped like budding yeasts, can be generated by immersing certain hyphal species in permeating media of high osmotic pressure (Kaminskyj *et al.,* 1992; Money & Harold, 1992, 1993). Because those authors were studying turgor pressure, or lack thereof, their overall tendency was to credit the 'softness' of the walls to growth at reduced turgor, although Money & Harold also suggested nutritional causes, and eventually (1992) considered that to be the more convincing hypothesis.

We agree it is less than likely that reduced turgor is the immediate cause. If walls are assembled as discussed above, via the association and integration of endolytic and synthetic activities, of exocytosis of matrix compounds, enzymes and their substrates, and of cross-linking mechanisms, there are many opportunities for the fabric of wall generation to become frayed. Even if one has too little insight to divine how the reduced turgor pressure experiments can affect one or more of the associated activities, or can affect their integration, there are other examples to consider. For instance, Vraná (1983a, b) has shown how some chemostat-grown fission yeasts incubated at low dilution rate, and thus under nutrient limitation, divide asymmetrically, yielding sibs of unequal volume. Although the relationships were not simple, the shorter sib was changed from typically sausage-like to typically budding-yeast like (ellipsoidal). Profound though these changes of shape were, they were almost trivial compared with the 'round-bottom-flask' fission yeast deformities induced by aculeacin A (Fig. 7; Miyata, Kitamura & Miyata, 1980). It was shown that α-glucan synthesis carried on apparently normally in the absence of β-glucan synthesis (Miyata, Kanbe & Tanaka, 1985), and finally it became clear that large volume asymmetries at cell division, long after removing the drug, were simply consequences of the deformities (Miyata, Miyata & Johnson, 1986a, b). The simplest explanation was that imbalanced glucan synthesis led to aberrant cell shapes, and that all else was a consequence of the cells following their usual rules for extensile growth and division while coping with the deformed cell shapes that acted as unusual templates for further morphogenesis.

In all these experiments, hyphae (long in *Achlya* (Money & Harold, 1993) or short in *S. pombe* (Vran, 1983a, b)) became prolate spheroids or worse (Miyata *et al.,* 1985). Hence we argue that the immediate cause in all might well have been an unbalanced synthesis or wall assembly that was only casually, *not causally,* correlated with reduced turgor pressure in the case of *Achlya* but not at all in the case of *S. pombe,* and might have been nutritionally driven as suggested by Money & Harold (1992). The argument seems reinforced by the observation that raising the external osmotic pressure with non-permeating solute reduced the turgor pressure but did not lead to softened walls (Money & Harold, 1993). This ingenious experiment divorced softening from lowered turgor pressure, allowing the nutritional or imbalance rationalizations to come to the fore.

Bent hyphae

That hyphae bend, hence grow in patterns that are not monaxial, must be evident to all mycologists (Chapter 3). Possibly the loveliest illustration of this is Buller's drawing of a young monosporous colony of *Coprinus sterquilinus* (Buller, 1924), which has been used to good effect to epitomise the fungal growth pattern (e.g. Trinci, 1978 (as Fig. 8.13), and as a jacket illustration by Burnett & Trinci, 1979). Our concern here is not with bending as a response to the environment, but rather with the mechanism for changing the axis. The possibilities seem to lie in but two categories: accidentally changed axis and deliberately changed axis.

If at any time, extension ceases and then is resumed at the same growing point, there is opportunity for an accidentally changed axis. Referring to the discussions of actin as a virtual memory above, let us note that at dispersal of most of the tip actin plaques associated with every cessation of extension or at re-aggregation of the tip actin plaques with every re-initiation of extension, there is the possibility of asymmetric dispersal or re-aggregation. Indeed, the wonder is that this is not the usual pattern. Fission yeast grown in chemostats at 20°C showed bends at old fission scars from which extension was initiated (Fig. 8). Because it is apparent that some of the old fission scar has disappeared, we assume that this extension was not monaxial from its very inception, and that the autolytic elements of extension have removed evidence of fission where the scar has disappeared. A probably comparable, but certainly rare, example of a bent fission yeast cell from normal batch culture is also shown (Fig. 9). The bent hyphae of *Coprinus* in Buller's picture may have developed in much the same manner as these. On the other hand it is possible that those bends occurred via a mechanism for deliberately changed axis, of which the most intimately studied examples are those driven by forbidden sites for extension: initiation of buds on haploid budding yeasts and of all buds of diploid yeasts after the first (discussed above). All of these buds are generated off the

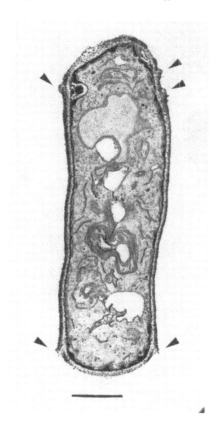

Fig. 8. Transmission electron micrograph of deformed fission yeast cell (*Schizosaccharomyces pombe*) after incubation in chemostat at 20°C. Arrow heads indicate fission scars, three on right side of micrograph and two on left. See text. Bar = 1 μm. (Micrographs by B. Y. Yoo of chemostat culture prepared by Drs T. W. James, R. Bohman, and the late Dr I. J. McDonald).

original primary axis, now forbidden, of the paternal cell. The reader is referred to Chant & Herskowitz (1991) for mechanisms of decision (but warned to beware their idiosyncratic conventions for axiallity).

Central controls

The extent to which extension is a basically autonomous process, or to which it is directly and intimately under central management remains unanswered and is beyond our scope. We have deliberately introduced the budding yeasts as specific examples of rather direct control by a fungal

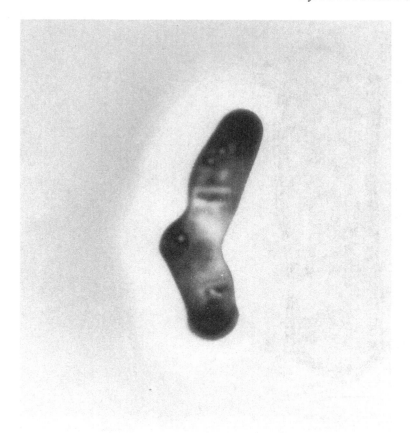

Fig. 9. Phase-contrast photomicrograph of fission yeast cell (*Schizosaccharomyces pombe*), quite comparable to the cell of Fig. 8. This unusual, bent cell was from a batch culture incubated at 30°C in 2% malt extract broth (Oxoid). (Courtesy of Ms Sabrina Piombo).

genome, but which can be generalized only after very intensive study of fungal hyphae.

The future

Those of us who are generalists see fungi as typical eukaryotes. However, the nature of their walls sets them apart. Hopes for anti-fungal chemotherapy are more likely to be fulfilled if drugs are designed against such unique aspects of their biology. The important qualitative differences are in the autolytic and the re-synthetic steps, and here is where chemotherapeutic targets must be waiting for exploitation.

The new hybrid model stresses the need for controlled endolytic activity and resynthesis of polysaccharide at the tip (from the old endolytic model) and the need for the post-hardening and cross-linking activities (from the steady-state model). The new hybrid model utilizes the verifiable strengths of both models and avoids the falsifiable weaknesses of both. However, the new hybrid model seems to have no application to intussusceptional growth nor to non-accretional growth. Although the paradigms seem to be in order, suggestions of basic models for these forms of hyphal expansion must wait until our understanding increases. Past studies of fungal morphogenesis have been very exciting; the future seems to hold equal or even greater excitement.

Acknowledgements

We thank Prof. Dr J. G. H. Wessels for generous discussion and Ms Sabrina Piombo for providing unpublished data. Drs R. Haskins and I. B. Heath have generously made helpful responses to a variety of questions. BFJ is grateful for the hospitality of Dr John Webb, Carleton University. BFJ & BYY are recipients of NSERC grants.

References

Alexopoulos, C. J. (1962). *Introductory Mycology*, 2nd. edn. John Wiley: New York.

Aronson, J. M. (1981). Cell wall chemistry, ultrastructure, and metabolism. In *Biology of Conidial Fungi*. Vol.2. (ed. G. T. Cole & B. Kendrick), pp. 459–507. Academic Press: New York.

Aronson, J. M. & Preston, R. D. (1960). The microfibrillar structure of the cell walls of the filamentous fungus, *Allomyces*. *Journal of Biophysical and Biochemical Cytology* **8**, 247–256.

Ashton, M.-L. & Moens, P. B. (1982). Light and electron microscopy of conjugation in the yeast, *Schizosaccharomyces octosporus*. *Canadian Journal of Microbiology* **28**, 1059–1077.

Barras, D. R. (1969). An endo-β-1,3-glucanase from *Schizosaccharomyces pombe*. *Antonie van Leeuwenhoek* **35**, Supplement, I17–I18.

Barras, D. R. (1972). A β-glucan endo-hydrolase from *Schizosaccharomyces pombe* and its role in cell wall growth. *Antonie van Leeuwenhoek* **38**, 65–80.

Bartnicki-Garcia, S. (1973). Fundamental aspects of hyphal morphogenesis. In *Microbial Differentiation* (ed. J. M. Ashworth & J. E. Smith), pp. 245–267. Cambridge University Press: Cambridge, U.K.

Bartnicki-Garcia, S. & Lippman, E. (1969). Fungal morphogenesis: cell wall construction in *Mucor rouxii. Science* **165**, 302–304.

Bartnicki-Garcia, S. & Lippman, E. (1972). The bursting tendency of hyphal tips of fungi: presumptive evidence for a delicate balance between wall synthesis and wall lysis in apical growth. *Journal of General Microbiology* **73**, 487–500.

Beran, K., Streiblová E. & Lieblová, J. (1969). On the concept of the population of the yeast *Saccharomyces cerevisiae. Proceedings of the 2nd Symposium on Yeasts*, pp. 353–363.

Bergman, K, Burke, P. V., Cerdá-Olmedo, E., David, C. N., Delbrück, M., Foster, K. W., Goodell, E. W., Heisenberg, M., Meissner, G., Zalokar, M., Dennison, D. S. & Shropshire, W. Jr. (1969). *Phycomyces. Bacteriological Reviews* **33**, 99–157.

Biely, P., Kovarik, J. & Bauer, S. (1973). Lysis of *Saccharomyces cerevisiae* with 2-deoxy-2-fluoro-D-glucose, an inhibitor of the cell wall glucan synthesis. *Journal of Bacteriology* **115**, 1108–1120.

Bretscher, A., Drees, B., Harsay, E., Schott, D. & Wang, T. (1994). What are the basic functions of microfilaments? Insights from studies in budding yeast. *Cell* **126**, 821–825.

Buller, A. H. R. (1924). *Researches on Fungi*, vol. 3. Longmans, Green & Co.: London.

Burnett, J. H. & Trinci, A. P. J. (1979). *Fungal Walls and Hyphal Growth.* Cambridge University Press: Cambridge, U.K.

Calleja, G. B., Johnson, B. F. & Yoo, B. Y. (1981). The cell wall as sex organelle in fission yeast. In *Sexual Interactions in Eukaryotic Microbes* (ed. D. H. O'Day & P. A. Horgen), pp. 225–259. Academic Press: New York.

Calleja, G. B., Yoo, B. Y. & Johnson, B. F. (1977a). Fusion and erosion of cell walls during conjugation in the fission yeast (*Schizosaccharomyces pombe*). *Journal of Cell Science* **25**, 139–155.

Calleja, G. B., Yoo, B. Y. & Johnson, B. F. (1977b). Conjugation-induced lysis of *Schizosaccharomyces pombe. Journal of Bacteriology* **130**, 512–515.

Campbell, A. (1957). Synchronization of cell division. *Bacteriological Reviews* **21**, 263–272.

Castle, E. S. (1953). Problems of oriented growth and structure in *Phycomyces. Quarterly Review of Biology* **28**, 364–372.

Chant, J. & Herskowitz, I. (1991). Genetic control of bud site selection in yeast by a set of gene products that constitute a morphogenetic pathway. *Cell* **65**, 1203–1212.

Cid, V. J., Durán, A., del Rey, F., Snyder, M. P., Nombela, C. & Snchez, M. (1995). Molecular basis of cell integrity and morphogenesis in *Saccharomyces cerevisiae. Microbiological Reviews* **59**, 345–386.

Cortat, M., Matile, P. & Wiemken, A. (1972). Isolation of glucanase-containing vesicles from budding yeast. *Archiv für Mikrobiologie* **82**, 189–205.

Engelhardt, W. A. (1942). Enzymatic and mechanical properties of muscle proteins. *Yale Journal of Biology and Medicine* **15**, 21–38.

Fèvre, M. (1979). Glucanases, glucan synthases and wall growth in *Saprolegnia monoica*. In *Fungal Walls and Hyphal Growth* (ed. J. H. Burnett & A. P. J. Trinci), pp. 225–263. Cambridge University Press: Cambridge, U.K.

Field, C. & Schekman, R. (1980). Localized secretion of acid phosphatase reflects the pattern of cell surface growth in *Saccharomyces cerevisiae. Journal of Cell Biology* **86**, 123–128.

Fleet, G. H. & Phaff, H. J. (1974). Glucanases in *Schizosaccharomyces*. Isolation and properties of the cell wall-associated β-(1->3)-glucanases. *Journal of Biological Chemistry* **249**, 1717–1728.

Flegel, T.W. (1978). Difference in generation times for mother and daughter cells in yeasts. *Canadian Journal of Microbiology* **24**, 827–833.

Gooday, G. W. (1977). Biosynthesis of the fungal wall – mechanisms and implications. *Journal of General Microbiology* **99**, 1–11.

Gooday, G. W. (1983). The hyphal tip. In *Fungal Differentiation: A Contemporary Synthesis* (ed. J. E. Smith), pp. 315–356. Marcel Dekker: New York.

Green, P. B. (1969). Cell morphogenesis. *Annual Review of Plant Physiology* **20**, 365–394.

Hartwell, L. H. & Unger, M. W. (1977). Unequal division in *Saccharomyces cerevisiae* and its implications for the control of cell division. *Journal of Cell Biology* **75**, 422–435.

Haskins, R. H. (1967). Backward-directed fungal hyphal branching. *Nature* **214**, 517–518.

Heath, I. B. (ed.) (1990). *Tip Growth in Plant and Fungal Cells.* Academic Press: San Diego.

Heilbrunn, L. V. (1952) *An Outline of General Physiology,* 3rd edn. W. B. Saunders: Philadelphia.

Hunsley, D. & Burnett, J. H. (1970). The ultrastructural architecture of the walls of some hyphal fungi. *Journal of General Microbiology* **62**, 203–218.

Inoué, S. (1989). Imaging of unresolved objects, superresolution, and precision of distance measurement with video microscopy. *Methods in Cell Biology* **30**, 85–112.

Jackson, S. L. & Heath, I. B. (1990). Evidence that actin reinforces the extensible hyphal apex of the oomycete *Saprolegnia ferax. Protoplasma* **157**, 144–153.

James, T. W., Hemond, P., Czer, G. & Bohman, R. (1975). Parametric analysis of volume distributions of *Schizosaccharomyces pombe* and other cells. *Experimental Cell Research* **94**, 267–276.

Johnson, B. F. (1965). Autoradiographic analysis of regional cell wall growth of yeasts. *Schizosaccharomyces pombe. Experimental Cell Research* **39**, 613–624.

Johnson, B. F. (1967). Growth of the fission yeast, *Schizosaccharomyces pombe,* with late, eccentric, lytic fission in an unbalanced medium. *Journal of Bacteriology* **94**, 192–195.

Johnson, B. F. (1968a). Lysis of yeast cell walls induced by 2-deoxyglucose at their sites of glucan synthesis. *Journal of Bacteriology* **95**, 1169–1172.

Johnson, B. F. (1968b). Dissolution of yeast glucan induced by 2-deoxyglucose. *Experimental Cell Research* **50**, 692–694.

Johnson, B. F. (1968c). Morphometric analysis of yeast cells. II. Cell size of *Schizosaccharomyces pombe* during the growth cycle. *Experimental Cell Research* **49**, 59–68.

Johnson, B. F., Calleja, G. B. & Yoo, B. Y. (1977). A model for controlled autolysis during differential morphogenesis of fission yeast. In *Eukaryotic Microbes as Model Developmental Systems* (ed. D. H. O'Day & P. A. Horgen), pp. 212–229. Marcel Dekker: New York.

Johnson, B. F., Calleja, G. B., Yoo, B. Y., Zuker, M. & McDonald, I. J. (1982). Cell division: key to cellular morphogenesis in the fission yeast, *Schizosaccharomyces. International Review of Cytology* **75**, 167–208.

Johnson, B. F. & Gibson, E. J. (1966a). Autoradiographic analysis of regional cell wall growth of yeasts. II. *Pichia farinosa*. *Experimental Cell Research* **41**, 297–306.

Johnson, B. F. & Gibson, E. J. (1966b). Autoradiographic analysis of regional cell wall growth of yeasts. III. *Saccharomyces cerevisiae*. *Experimental Cell Research* **41**, 580–591.

Johnson, B. F. & Lu, C. (1975). Morphometric analysis of yeast cells. IV. Increase of the cylindrical diameter of *Schizosaccharomyces pombe* during the cell cycle. *Experimental Cell Research* **95**, 154–158.

Johnson, B. F., Lu, C. & Brandwein S. (1974). Morphometric analysis of yeast cells. III. Size distribution of 2-deoxyglucose-induced lysing *Schizosaccharomyces pombe* cells and their sites of lysis. *Canadian Journal of Genetics and Cytology* **16**, 593–598.

Johnson, B. F. & McDonald, I. J. (1983). Cell division: a separable cellular sub-cycle in the fission yeast *Schizosaccharomyces pombe*. *Journal of General Microbiology* **129**, 3411–3419.

Johnson, B. F., Miyata, M. & Miyata, H. (1989). Morphogenesis of fission yeasts. In *Molecular Biology of the Fission Yeast* (ed. A. Nasim, P. R. Young & B. F. Johnson), pp. 331–366. Academic Press: San Diego.

Johnson, B. F. & Rupert, C. M. (1967). Cellular growth rates of the fission yeast, *Schizosaccharomyces pombe,* and variable sensitivity to 2-deoxyglucose. *Experimental Cell Research* **48**, 618–620.

Johnson, B. F., Yoo, B. Y. & Calleja, G. B. (1995). Smashed fission yeast walls. Structural discontinuities related to wall growth. *Cell Biophysics* **26**, 57–75.

Johnston, G. C., Pringle, J. R. & Hartwell, L. H. (1977). Coordination of growth with cell division in the yeast *Saccharomyces cerevisiae*. *Experimental Cell Research* **105**, 79–98.

Kaminskyj, S. G. W., Garrill, A. & Heath, I. B. (1992). The relation between turgor and tip growth in *Saprolegnia ferax*: turgor is necessary, but not sufficient to explain apical extension rates. *Experimental Mycology* **16**, 64–75.

Klis, F. M. (1994). Review: Cell wall assembly in yeast. *Yeast* **10**, 851–869.

Kreger-van Rij, N. J. W. (1977). Electron microscopy of sporulation in *Schwanniomyces alluvius*. *Antonie van Leeuwenhoek* **43**, 55–64.

Kritzman, G., Chet, I. & Henis, Y. (1978). Localization of β-(1,3)-glucanase in the mycelium of *Sclerotium rolfsii*. *Journal of Bacteriology* **134**, 470–475.

López-Franco, R., Bartnicki-Garcia, S. & Bracker, C. E. (1994). Pulsed growth of fungal hyphal tips. *Proceedings of the National Academy of Science, U.S.A.* **91**, 12228–12232.

MacKay, V. & Manney, T. R. (1974). Mutations affecting sexual conjugation and related processes in *Saccharomyces cerevisiae*. I. Isolation and phenotypic characterization of nonmating mutants. *Genetics* **76**, 255–271.

Marks, J. & Hyams, J. S. (1985). Localization of F-actin through the cell division cycle of *Schizosaccharomyces pombe*. *European Journal of Cell Biology* **39**, 27–32.

Marks, J., Hagan, I. M. & Hyams, J. S. (1987). Spatial association of F-actin with growth polarity and septation in the fission yeast *Schizosaccharomyces pombe*. *Special Publication of the Society for General Microbiology* **23**, 119–135.

May J. W. & Mitchison, J. M. (1986). Length growth in fission yeast cells measured by two novel techniques. *Nature* **322**, 752–754.

McMacken, R., Silver, L. & Georgopoulos, C. (1987). DNA replication. In *Escherichia coli and Salmonella typhimurium. Cellular and Molecular Biology,* vol 1 (ed. F. C. Neidhardt, J. L. Ingraham, K. B. Low, B. Magasanik, M. Schaechter & H. E. Umbarger), pp. 564–612. ASM: Washington.

Megnet, R. (1965). Effect of 2-deoxyglucose on *Schizosaccharomyces pombe. Journal of Bacteriology* **90,** 1032–1035.

Mitchison, J. M. (1957). The growth of single cells. I. *Schizosaccharomyces pombe. Experimental Cell Research* **13,** 244–262.

Mitchison, J. M. & Nurse, P. (1985). Growth in cell length in the fission yeast *Schizosaccharomyces pombe. Journal of Cell Science* **75,** 357–376.

Miyata, H., Miyata, M. & Johnson, B. F. (1986). Patterns of extension growth of the fission yeast, *Schizosaccharomyces pombe. Canadian Journal of Microbiology* **32,** 528–530.

Miyata, H., Miyata, M. & Johnson, B. F. (1988). Pseudo-exponential growth in length of the fission yeast, *Schizosaccharomyces pombe. Canadian Journal of Microbiology* **34,** 1338–1343.

Miyata, H., Miyata, M. & Johnson, B. F. (1990). Pattern of end growth of the fission yeast *Schizosaccharomyces pombe. Canadian Journal of Microbiology* **36,** 390–394.

Miyata, M., Kanbe, T. & Tanaka, K. (1985). Morphological alterations of the fission yeast *Schizosaccharomyces pombe* in the presence of aculeacin A: spherical wall formation. *Journal of General Microbiology* **131,** 611–621.

Miyata, M., Kitamura, J. & Miyata H. (1980). Lysis of growing fission-yeast cells induced by aculeacin A, a new antifungal antibiotic. *Archives of Microbiology* **127,** 11–16.

Miyata, M. & Miyata H. (1978). Relationship between extracellular enzymes and cell growth during the cell cycle of the fission yeast *Schizosaccharomyces pombe*: acid phosphatase. *Journal of Bacteriology* **136,** 558–564.

Miyata, M., Miyata H. & Johnson, B. F. (1986a). Asymmetric location of the septum in morphologically altered cells of the fission yeast *Schizosaccharomyces pombe. Journal of General Microbiology* **132,** 883–891.

Miyata, M., Miyata H. & Johnson, B. F. (1986b). Establishment of septum orientation in a morphologically altered fission yeast *Schizosaccharomyces pombe. Journal of General Microbiology* **132,** 2535–2540.

Money, N. P. & Harold, F. M. (1992). Extension growth of the water mold *Achlya*: interplay of turgor and wall strength. *Proceedings of the National Academy of Science,* U.S.A. **89,** 4245–4249.

Money, N. P. & Harold, F. M. (1993). Two water molds can grow without measurable turgor pressure. *Planta* **190,** 426–430.

Moore, D. (1969). Effect of 2-deoxy-D-glucose on mycelial growth of filamentous fungi. *Transactions of the British Mycological Society* **53,** 139–141.

Moore, D. (1981). Effects of hexose analogues on fungi: mechanisms of inhibition and of resistance. *New Phytologist* **87,** 487–515.

Mulholland, J., Preuss, D., Moon, A., Wong, A., Drubin, D. & Botstein, D. (1994). Ultrastructure of the yeast actin cytoskeleton and its association with the plasma membrane. *Journal of Cell Biology* **125,** 381–391.

Novick, P., Field, C. & Schekman, R. (1980). Identification of 23 complementation groups required for post-translational events in the yeast secretory pathway. *Cell* **21**, 205–215.

Park, D. & Robinson, P.M. (1966). Aspects of hyphal morphogenesis in fungi. In *Trends in Plant Morphogenesis* (ed. E. G. Cutter), pp. 27–44. Longmans, Green: London.

Popper, K. R. (1983) *Realism and the Aim of Science*. Rowman & Littlefield: Totowa, NJ, U.S.A.

Prosser, J. I. & Trinci, A. P. J. (1979). A model for hyphal growth and branching. *Journal of General Microbiology* **111**, 153–164.

Ray, P. M., Green, P. B. & Cleland, R. (1972). Role of turgor in plant cell growth. *Nature* **239**, 163–164.

Reinhardt, M. O. (1892). Das Wachsthum der Pilzhyphen. Ein Beitrag zur Kenntniss des Flächenwachsthums vegetabilischer Zellmembranen. *Jahrbücher für wissenschaftliche Botanik* **23**, 479–566.

Robertson, N. F. (1965a). The fungal hypha. *Transactions of the British Mycological Society* **48**, 1–8.

Robertson, N. F. (1965b). The mechanism of cellular extension and branching. In *The Fungi,* Vol. 1 *The Fungal Cell,* (ed. G. C. Ainsworth & A. S. Sussman), pp. 613–23. Academic Press: New York.

Robertson, N. F. (1968). The growth process in fungi. *Annual Review of Phytopathology* **6**, 115–136.

Robinow, C. F. (1963). Observations on cell growth, mitosis, and division in the fungus *Basidiobolus ranarum*. *The Journal of Cell Biology* **17**, 123–152.

Robinow, C. F. & Hyams, J. S. (1989). General cytology of fission yeasts. In *Molecular Biology of the Fission Yeast* (ed. A. Nasim, P. R. Young & B. F. Johnson), pp. 273–330. Academic Press: San Diego.

Robinow, C. F. & Johnson, B. F. (1991). Yeast cytology: an overview. In *The Yeasts*, vol. **4**, 2nd edn (ed. A. H. Rose & J. S. Harrison), pp. 7–120. Academic Press: London.

Rosenberger, R. F. (1979). Endogenous lytic enzymes and wall metabolism. In *Fungal Walls and Hyphal Growth* (ed. J. H. Burnett & A. P. J. Trinci), pp. 265–277. Cambridge University Press: Cambridge, U.K.

Russell, I., Garrison, I. F. & Stewart, G. G. (1973). Studies on the formation of spheroplasts from stationary phase cells of *Saccharomyces cerevisiae*. *Journal of the Institute of Brewing* **79**, 48–54.

Salmon, T., Walker, R. A. & Pryer, N. K. (1989). Advances in microscopy – part III. Video-enhanced differential interference contrast light microscopy. *BioTechniques* **7**, 624–633.

Sipiczki, M., Grallert, B. & Miklos, I. (1993). Mycelial and syncytial growth in *Schizosaccharomyces pombe* induced by novel septation mutations. *Journal of Cell Science* **104**, 485–493.

Smith, J. E. & Berry, D. R. (eds) (1978). *The Filamentous Fungi.* Vol.3 *Developmental Mycology.* Edward Arnold: London.

Staebell, M. & Soll, D. R. (1985). Temporal and spatial differences in cell wall expansion during bud and mycelium formation in *Candida albicans*. *Journal of General Microbiology* **131**, 1467–1480.

Streiblov, E. (1970). Study of scar formation in the life cycle of heterothallic *Saccharomyces cerevisiae*. *Canadian Journal of Microbiology* **16**, 827–831.

Thomas, D. des S. & Mullins, J. T. (1967). Role of enzymatic wall-softening in plant morphogenesis: hormonal induction in *Achlya*. *Science* **156**, 84–85.

Townsend, G. F. & Lindegren, C. C. (1954). Characteristic growth patterns of the different members of a polyploid series of *Saccharomyces*. *Journal of*

Bacteriology **67**, 480–483.

Trinci, A. P. J. (1978). The duplication cycle and vegetative development in moulds. In *The Filamentous Fungi*, **Vol.3** *Developmental Mycology* (ed. J. E. Smith & D. R. Berry), pp. 132–163. Edward Arnold: London.

Trinci, A. P. J. (1979). The duplication cycle and branching in fungi. In *Fungal Walls and Hyphal Growth* (ed. J. H. Burnett & A. P. J. Trinci), pp. 319–358. Cambridge University Press: Cambridge, U.K.

Trinci, A. P. J. & Cutter, E. G. (1986). Growth and form in lower plants and the occurrence of meristems. *Philosophical Transactions of the Royal Society of London* **B313**, 95–113.

Vraná, D. (1983a). Morphological properties of *Schizosaccharomyces pombe* in a continuous culture. *Folia Microbiologica* **28**, 414–419.

Vraná, D. (1983b). The fission yeast *Schizosaccharomyces pombe* in continuous culture. *Biotechnology and Bioengineering* **25**, 1989–1994.

Wessels, J. G. H. (1993). Wall growth, protein excretion and morphogenesis in fungi. *New Phytologist* **123**, 397–413.

Wessels, J. G. H. & Sietsma, J. H. (1981). Fungal cell walls: a survey. In *Encyclopedia of Plant Physiology New Series* **13 B,** *Plant Carbohydrates II* (ed. W. Tanner & F. A. Loewus), pp. 352–394. Springer-Verlag: Berlin.

Winge, Ö. (1935). On haplophase and diplophase in some Saccharomycetes. *Comptes Rendus des travaux du Laboratoire Carlsberg (Série Physiologique)* **21**, 77-115.

Chapter 3

Pattern formation and development of the fungal mycelium

KEITH K. KLEIN

Summary

The development of distinct patterns by fungal mycelia is a direct consequence of growth. In one case, mycelia develop from an undifferentiated inoculum into complex structures with distinct zones of differentiated cells forming a pattern of concentric rings on solid media. Models have been developed which can describe the formation of these zones, and which form the basis for further elaboration to include the development of significant heterogeneities in the mycelium. Among these heterogeneities are the presence of circular ring patterns, spirals, and asymmetries. These in turn help to explain the formation of linear and circular differentiated structures such as rhizomorphs, cords and sclerotia. The appearance of some of these structures has been shown to be under both internal and environmental control, with the rhythmic circadian patterns of hyphal growth the best understood at both biochemical and genetical levels. Mycelial pattern formation has great significance for the growth of fungi in liquid culture, especially in the production of differentiated pelleted forms. An understanding of the role of fermentation procedures in the development of mycelial patterns and on the selection of morphological mutants by culturing mycelia has begun to be elaborated.

Introduction

In nature fungal mycelia are rarely seen and seldom looked for. Yet these hidden parts of fungi can grow to enormous mass spreading beneath vast areas and might be among the most long lived organic structures on earth (Smith, Bruhn & Anderson, 1992). The mycelial structure may take many forms, including rhizomorphs, cords, fans, mycorrhizal sheaths about plant roots, sclerotia, pseudosclerotia and the 'diffuse' or spreading form familiar to the laboratory culture of fungi (Rayner, 1991, provides a good review). It is the diffuse structure, often mistakenly called a 'colony', that is best known especially as it can be grown easily on an agar surface. As has been amply demonstrated, however, this mass of cells is far more than a collection of independent organisms. It is a self-interacting network of tubular cells, which form the vegetative part of the fungal thallus, performing the functions of nutrient acquisition, growth, dormancy, and migration and which forms the supporting structure for the development of fruiting stages. It performs these functions by the development of specialized mycelial elements mentioned above, all of which may be produced readily in laboratory culture, and which constitute the vegetative organs of the fungus (Rayner *et al.,* 1985). Since the mycelium is capable of a wide range of developmental responses to environmental factors using a very restricted repertoire of cell shapes (i.e. tubular), and fungi are easily adapted to growth in laboratory conditions, the fungal mycelium is an ideal candidate for the experimental study of the mechanisms of development.

The pattern displayed by a mycelium is formed by the spatial distribution of cells and cell types. It develops over time by the growth of individual cells which in their various modes of elongation, branching and fusion give rise to the different tissues and organs of fungi. Cells in the mycelium are not motile, nor are they able to effect rapid changes in their sizes and shapes. Thus the form which any mycelium takes is a result of growth of fungal cells under environmental conditions suitable for the particular species of fungus and is manifested in the physical appearance of the mycelium in question. Much of the growth pattern of mycelia may be understood as responses to stimuli from outside the organism but only as they affect the physiological parameters of growth intrinsic to the organism. Particular chemicals or certain wavelengths of light, physical properties such as grain structure or fluidity of the surrounding medium may serve as cues which provoke a particular kind of mycelial growth.

Before an understanding of the development of the fungal mycelium can be reached, the parameters of growth must first be defined. These will include the nutrient status of the organism, the physical conditions of its surroundings, its status with regard to mating behaviour, and its intrinsic growth habit. Models which incorporate these parameters in an analytical sense may prove to be particularly beneficial to the understanding of growth and developmental processes and may provide a useful basis for experimentation. Ideally, a model should provide insight into the processes under consideration in a quantifiable manner. Models for the growth of unicellular microorganisms have long been developed but these have not proved to be of equal utility in the description of fungal growth. Even though it is possible to culture mycelia by means similar to bacterial methods, there are profound differences in the kind of growth which occurs (Koch, 1975). These differences have important consequences for the use of filamentous forms in industrial as well as laboratory fermentations.

The growth and development of mycelia have been studied extensively, both as they occur in liquid media and on solid or semi-solid surfaces. The form of growth in liquid culture is related to, but not the same as, that of surface cultures. Aside from the question of dimensionality, there are the problems of shear forces, turbulence and diffusion to consider in any discussion of developmental differences produced by the liquid and solid culture conditions. It may be argued with some justification that for fungi of interest to the developmental biologist growth on solid or semi-solid media is more like natural conditions than growth in liquid culture. Accordingly, most of this review will be devoted to discussion of work done on fungi grown on surfaces. But due to the enormous commercial potential of liquid culture methods, or fermentations using filamentous fungi, some mention will be made of work done on development of liquid-grown mycelia.

Pattern formation in surface cultures: diffuse mycelia

Mycelia, unlike single cell organisms, form as a community of hyphae without separation of daughter cells. The initial cell may be a sexual or asexual spore or a fragment of pre-existing mycelia but in the case of a filamentous free-living mycelium the form of growth from any of these inocula on artificial medium is essentially the same. Mycelial development proceeds by the growth of tubular cells at their apices and by branching eventually producing a 'colony' of interconnected cells (Chapter 2). For any growing mycelium under laboratory conditions one may define several growth parameters related to cellular increase: rate of elongation at the tip, site of branch initiation, angle formed by branches with the original hyphal axis and frequency of branch initiation. Control of these processes is both genetic and environmental (Trinci, 1973a), and the processes have profound effects upon the forms the mycelia take. Among the characters affected are: hyphal density, rate of utilization of surface resources, and potential for further development.

The inoculum initially enters a lag, or swelling, phase and then the cells of the inoculum proceed to grow in an exponential ('logarithmic') form. This form of growth may be present for only a brief time (Trinci, 1970; 1971;

Anderson & Deppe, 1976). This exponential growth can be shown by concomitant increase in mass, total hyphal length, number of hyphal tips, number of nuclei, wet weight, or other such measures (Trinci, 1973b; Fiddy & Trinci, 1976a & b; Anderson & Deppe, 1976). After this period of exponential growth, the rate of growth of each hyphal tip declines until it reaches a steady-state (Trinci, 1974). In the steady-state the hyphae increase in length in proportion to the fungal growth unit. Being a unit length of hyphae associated with each hyphal tip, a unit remains more or less constant as the mycelium ages (Trinci, 1973b). Rates of elongation of various species and strains of fungi are associated in a linear manner with the absolute size of this growth unit such that the rate of elongation is equivalent to the intrinsic growth rate multiplied by the size of the growth unit (Bull & Trinci, 1977). Differences may occur between two-dimensional (surface) growth and growth in three dimensions in liquid medium in the distribution of the growing zone about the periphery of the mycelium (Koch, 1975). Mycelia grown in submerged culture may have greater variability in the dimensions of the growth zone and consequently in the size of the growth unit than the surface mycelia. The rate of hyphal elongation is in turn related to the rate of branch initiation in a direct manner, suggesting that branch initiation is controlled by the capacity of the hypha to add mass, i.e. its specific growth rate (Katz, Goldstein & Rosenberger, 1972).

Differentiation of the mycelium

After the cessation of exponential growth, the mycelium may begin to show signs of differentiation of its cells into different hyphal forms. In some species these are seen as wide leading hyphae and narrower secondary hyphae. Leading hyphae continue to grow either at the rate established by the undifferentiated mycelium or they may increase in growth rate and give rise to the secondary hyphae by branching behind the growth front (Trinci, 1974; Robinson & Smith, 1980; Smith & Robinson, 1980; Gow & Gooday, 1982). These secondary hyphae grow at a slower rate (Steele & Trinci, 1975). It is clear that occasionally new primary hyphae must arise, otherwise the mycelium will take on a markedly asymmetrical form. Presumably, these arise from secondary hyphae by an increase in hyphal diameter. In the development of *Neurospora crassa* mycelia, the linear extension rate of the leading hyphae increases during differentiation as the diameters of the leading hyphae also increase. Timing of this event suggests that nutrient concentration gradients are not responsible for the observed changes.

Rather, the production and translocation of a morphogenetic substance or staling compound by the older, nutrient-limited hyphae near the centre of the mycelium is implied (McLean & Prosser, 1987). The mycelium at this time is differentiated into a pattern of concentric zones: the outermost growth zone, a zone of cellular proliferation and anastomoses, and an innermost zone of quiescence, which shows signs of cells autolysing and becoming vacuolated (Yamagita & Kogane, 1962). Depending upon environmental conditions, there may be a fruiting zone in which proliferation of the aerial structures into fruit bodies occurs. As the mycelium ages, the innermost regions become senescent, and the cells are highly vacuolated with their cytoplasm presumably exported in the direction of the growing front (Edelstein, 1982).

The growing front is where the majority of the new cells being added to the mycelium are to be found. Attempts to model this growth have established a good fit with a logistic type of growth curve (Trinci, 1973a). The logistic model is in good agreement with the patterns developed by mycelia with a uniform structure of concentric zones progressively grading into each other as one examines the mycelium along a radius. One problem of the logistic model is its failure to account for hyphal events such as growth only at hyphal tips, branching of hyphae and cell death. This difficulty has been addressed by the development of more comprehensive models which provide a more accurate description of growth of mycelia (Edelstein, 1982). These models typically include reaction-diffusion mechanisms which involve intercellular signalling as a prerequisite in the process of branch initiation. Such models have been used extensively in the explanation of developmental phenomena of a vast array of organisms other than fungi (Meinhardt, 1984) and are based on mathematical deductions of chemical reactions under certain limiting conditions (Turing, 1952). Recent work has elaborated upon these models by demonstrating that there must be an avoidance signal which directs growth of hyphal branches away from the axis of the parent hyphae in *Mucor* and other species (Indermitte, Liebling & Clemençon, 1994). The distinct avoidance reaction of growing hyphae is seen when hyphal tips are placed near each other (Hutchinson *et al.*, 1980), although the pattern of very early growth of these mycelia can be described by a stochastic model, implying significant random elements in the process (Kotov & Reshetnikov, 1990). In computer simulations of branching and growth, models which do not incorporate an avoidance mechanism do not produce circularly symmetrical mycelia. Such models support the hypothesis that there must be an inhibitor substance to prevent hyphae from coming into close contact, at least at the growing margins of circular mycelia. Hyphae may show branch angles that have a characteristic value as a result of these avoidance reactions. It remains to be seen if the absence of such a mechanism either by mutational loss or as a result of physiological control alters the development of mycelia from single cells.

Large scale patterns of surface mycelia

Two major departures from the uniform model of growth are the existence of rhythmic patterns of cell proliferation such as the concentric rings of conidiation in *Neurospora crassa* (Sargent, Briggs & Woodward, 1966), and the appearance of significant spontaneous deviations from circular symmetry which are found in the mycelia of many basidiomycetes and in some ascomycetes grown under low nutrient conditions (Matsuura & Miyazima, 1993b & c). In both of these cases, the development of the mycelium produces macroscopic patterns which are discernable at the scale of the entire mycelium and reminiscent of the developmental patterns displayed by plants and animals. The relative simplicity of the mycelium compared to organisms in other kingdoms offers some hope that the physiological and genetic mechanisms which regulate macroscopic development might be amenable to dissection. Recent work has reinforced this hope, and much progress has been made in elucidating the biochemical nature of these processes (see below). As with growth in general, biochemical modelling has been instrumental in producing a framework which allows a more complete understanding of these patterns of mycelial development.

Circular and spiral patterns superimposed on mycelial growth

The occurrence of rhythmic patterns in the growth of mycelia presents an interesting case for both modellers and physiologists. These patterns are known for a wide variety of fungi and may appear as rings or spirals of denser hyphae with or without conidia (reviewed by Lysek, 1983). The pattern may be generated in a circadian rhythm, which can be entrained to cycles of light and darkness (Sargent, Briggs & Woodward, 1966) or the pattern may not be entrained to any known outside stimulus (Bourret, Lincoln & Carpenter, 1969; Sharland & Rayner, 1989). Of the two alternative mechanisms circadian control of the formation of patterns has been the most intensively studied, especially in *Neurospora crassa*. The development of these patterns proceeds by the growth of alternate bands of thin or sparse aerial hyphae and/or conidia. In *Neurospora* and some other genera, these rings of alternate sparse and rich conidiation or aerial hyphae may form under conditions of constant environment, demonstrating the endogenous nature of their control. Genetic control of the period and the occurrence of this behaviour has been demonstrated (Feldman, 1982; Dunlap, 1993). Molecular cloning and examination of the circadian period genes has revealed that at least two of these genes identified by mutational analysis are under transcriptional control (Loros, Denome & Dunlap, 1989; Loros & Dunlap, 1991). The evidence provided by these rings of growth implies a cyclic mechanism with a chemical basis involving some form of auto-regulation via a negative feedback mechanism. Such a mechanism has recently been demonstrated (Aronson *et al.*, 1994) at least for the circadian clock of *Neurospora*. The clock itself is susceptible to resetting by blue light, and at least one of the genes involved in clock regulation is under both external control from light and internal control via the feedback mechanism of the clock (Arpaia *et al.*, 1993). The intracellular nature of the circadian rhythm of *Neurospora* and its ability to be reset by external stimuli are in direct contrast to the rhythm displayed by fresh cultures of *Penicillium clavigerum* (Piskorz-Binczycka, 1991). In this fungus, the bands of dense conidiation appear on a 72 hour rhythm, which is not resettable. The pattern is subject to alteration by externally applied asparagine and avidin which inhibit pattern formation and biotin which shortens the period of banding, implying an endogenous chemical signal as a regulator of pattern development. There may be extracellular signals involved as well. Small effector molecules have been implicated in the development of conidia of *Aspergillus flavus*, and these molecules are diffusible between mycelia separated by dialysis membrane (Lee & Adams, 1994).

In the development of these patterns of dense and thin mycelia and conidia, one needs to distinguish between intracellular mechanisms and intercellular mechanisms. In the case of the circadian rhythms of *Neurospora*, and perhaps the longer rhythms of *Penicillium*, which are governed by intracellular events, the pattern which appears must be the same for all regions of the mycelium which are at the same age, or have been subjected to the same 'clock setting' events. In the case of circadian rhythms, it is the environment (in the form of light) which acts as the overall organizing stimulus for the mycelium. One expectation of patterns developed in this manner is that they will disappear in conditions where the environmental stimulus is absent and growth effectively separates the mycelium into distinct regions no longer sharing cytoplasm. This is indeed the case for *Neurospora* mycelia grown in constant light (or dark), where the rhythmic phase of various parts of the mycelium may differ significantly (Dharmananda & Feldman, 1979). In *Penicillium*, the situation is much the same in older cultures, with the disappearance of the pattern with time of culture. Extracellular mechanisms, on the other hand, should show more persistence, due to the longer distances over which diffusible regulators can have their effect. The appearance of spiral patterns in *Nectria* is an instance of a pattern which cannot be explained except by an extracellular mechanism. Such spiral patterns also occur during the aggregation of *Dictyostelium* amoebae (Newell, 1983) and are highly reminiscent of certain chemical reactions which are diffusion limited (Becker & Field, 1976).

Other circular patterns may form on developing mycelia, such as the development of a ring of sclerotial tissue on mycelia of *Sclerotium rolfsii* (Okon, Chet & Henis, 1972). Both a chemical induction of sclerotium formation and a time-dependence of the process have been identified, but models which can explain the phenomenon have not been entirely successful (Edelstein *et al.*, 1983). Models which incorporate an age- and growth-related structure can produce the circular pattern of this sclerotial development at the cost of having the zone of sclerotial formation move outward radially with time (Edelstein & Segel, 1983). Clearly, the structure does not move once formed, and the model may be useful only in predicting where a sclerotium might form given the appropriate external stimuli.

Development of asymmetries of growth in diffuse mycelia

In certain instances, the mycelium may not grow on artificial medium in a circularly symmetric form. As with the development of rhythmic patterns, this asymmetry of growth habit may also be a consequence of fluctuations in the levels of some metabolites. Recent data suggests that in some cases the occurrence of new primary or leading hyphae may be suppressed (Matsuura & Miyazima, 1993a; Klein, unpublished observations). Both external factors, such as low nutrient levels, the presence of inhibitory compounds, and endogenous physiological effects as with autoinhibitory mechanisms have been implicated. In some cases, the development of mycelial asymmetries may be limited to secondary mycelia, especially the dikaryons of basidiomycetes (Sen, 1990; Fries & Sun, 1992). In any case, it is the development of asymmetry in a uniform environment which needs to be explained. Only mechanisms intrinsic to the mycelia themselves can produce the observed asymmetries under these conditions. The mechanism remains unknown, but some models would indicate that asymmetries of form can arise by chemical means in a reaction-diffusion system (Meinhardt, 1976). These models, which are all based on the work of Turing (1952) have several features which make them attractive in the explanation of growth and pattern formation at this stage. One of these is that there is significant non-linearity inherent in them, which can account for some observations of mycelia grown under laboratory conditions which are not otherwise explained. As with branch pattern formation, this departure from circular symmetry is best modelled by including a term which describes the accumulation of an auto-inhibitory compound (Meinhardt, 1976; mechanisms extensively reviewed in Murray, 1993). In some basidiomycetes, this is observed by an apparent antagonism of genetically identical mycelia growing into contact with each other. This occurs only when the mycelia commence growth sufficiently far apart to allow time for accumulation (Sen, 1990). A key feature of these asymmetries is that they are large with respect to single cells and compared to hyphal growth units. The effect of autoinhibition may also be seen in the growth kinetics of certain fungi with markedly irregular mycelia (Stone, Pinkerton & Johnson, 1994). Some means of communication across many hyphae is required, and extracellular effector molecules might well be that means.

The development of linear organs

A naturally occurring change from circularly symmetric, diffuse growth to asymmetry is that which occurs under conditions of nutrient starvation in certain cord-forming basidiomycetes (Rayner, 1991). In experiments where a mycelium is first grown on a wood block and then placed on a sterile, non-nutritive medium, mycelial cords are observed growing from the colonized block onto the medium (Rayner, 1991). These experiments parallel others where race tubes filled with soil are inoculated with mycelium at one end, and strands and cords are observed to grow out of the inoculum into the soil (Dowson, Rayner & Boddy, 1983). The formation of cords was observed to be greater in unsterilized soil than in soil sterilized with gamma irradiation (Thompson & Rayner, 1982). This would indicate that labile biological substances are at least partly involved in the induction of the linear growth form of these fungi. The linear organs produced by some fungi may be persistent for years in soil (Thompson & Boddy, 1983). They serve as the means of movement through the soil environment and promote the colonization of new substrates. They also serve as the means of translocation of large amounts of solute from a previously colonized substrate to the growing tips of the strands, cords or rhizoids (Jennings, 1987). The mycelium is able to reform itself into diffuse growth upon contact with suitable rich substrate. The system of cord formation from diffuse mycelium and diffuse mycelium growing from cords provides an experimentally manipulable developmental sequence which should provide rich insights into the mechanisms of pattern formation (Dowson *et al.*, 1986). In experiments to test the outgrowth patterns from colonized wood blocks onto non-nutritive medium, the outgrowth pattern of mycelial cords is at first broadly symmetric, although the level of organization is noticeable different from the pattern of diffuse mycelia consisting of multi-hyphal cords rather than a symmetrical distribution of individual hyphae (Dowson *et al.*, 1986). It would appear that a few of the cords form by some mechanism which 'chooses' among the diffuse hyphae to select leader hyphae which serve as the organisers of the newly formed cords. These cords will grow until either the resources of the mycelium are exhausted, or they contact an uncolonized patch of fresh medium or unused resources. The fungus will redirect its efforts to increasing the size of the cords which stretch between the old and new resource units and withdraw cytoplasm from the cords which did not contact new sources. The newly colonized resource will have within it a diffuse mycelium as an outgrowth from the cords (Dowson *et al.*, 1989). This ability to change morphology according to environmental signals is of course critical to the survival of the fungal mycelium, and even, perhaps, of the fungal species.

Of great importance to the development of mycelial systems is the ability to respond to other organisms in the environment. Demonstration of the effects of other organisms on soil-inhabiting fungi has been made and includes the production of pigments, dense mycelial networks, mycelial fans and cord formation (Rayner, Griffith & Wildman, 1994). In the production of these structures, diffusible substances from both self and non-self had inducing effects on these morphologies, as did the presence of metabolic inhibitors. The production of pattern in the development of linear structures from diffuse mycelia may be an expression of a form of inhibition within the mycelium. When a few selected hyphae grow into a linear cord or rhizomorph, it produces a suppression of the growth of surrounding hyphae. Such suppression, as mentioned above (Sen, 1990; Klein, unpublished) has been demonstrated for a limited number of cases with mycelia grown in culture. In some instances, the suppression may simply be due to an inadequate supply of nutrients such that only the most advanced hyphae in a growing front have access to an adequate supply (Matsuura & Miyazima, 1993c).

Practical considerations: fermenter operations

The growth and development of mycelia in liquid culture conditions is of enormous interest to those who must extract commercially important products from the culture of filamentous fungi (Reichl, King & Giles, 1992). The use of fungi as a source of commercially important products by liquid fermentation has become the technique most widely employed. Both batch and continuous culture are used in the production of fungal metabolites, and the form of mycelium which is found in any given system will have important consequences on the economics of production. Mycelial form can determine the yields, ease of extraction, speed of reaction and quality of product as well as the handling properties of the fluid medium itself (Olsvik *et al.*, 1993). Because the technique is similar in form to the use of bacteria and yeasts, the temptation has been to treat filamentous organisms as though they were similar in growth habit to these unicellular organisms. However, an efficient use of filamentous fungi in fermentations requires a precise knowledge of their growth and form in liquid culture. There are two generally recognized morphologies for submerged mycelia: filamentous and pelleted. Filamentous mycelia are open loosely aggregated forms which have a high proportion of living cells. Pellets on the other hand are much denser and may have significant necrotic tissue in their interiors (Wittler *et al.*, 1986). Factors which can cause a change in morphologies from one type to the other are oxygen saturation, shear stress in the medium, turnover rate, glucose levels, etc. (Weibe & Trinci, 1991; Weibe *et al.*, 1992).

Of the two morphologies which are found in mycelial liquid culture, filaments are the form preferred in the production of fungal metabolites. Filaments can range from single cells to very loose aggregations of cells and are distinguished by the ease with which a fluid medium may penetrate to all the cells of an aggregation. In this filamentous form all the cells of an aggregate are metabolically active, and there is very little to differentiate one cell from another. Pellets on the other hand, are composed of more-or-less spherical masses of many hyphae quite densely packed, and displaying an age structure with old, metabolically inactive cells at the centre of the pellet and younger more active cells at the surface. Growth of the two forms can be quite different, with the filamentous form growing at exponential rates in rich medium and the pelleted form growing with cube-root kinetics. In many systems, the production of filaments precedes the production of pellets, a process which has been modelled (Yang *et al.*, 1992a, 1992b; Viniegra-Gonzalez *et al.*, 1993).

The kinetics of growth in submerged cultures have been the subject of much study (Prosser, 1993). Models which would attempt to account for morphological development are of particular importance to the understanding of the fermentation process. These models can be analytic, as in the tubular reactor model of Aynsley, Ward & Wright (1990), which is of predictive value for assessing growth of hyphal tips, branching, and fragmentation rates, or descriptive as in certain fractal models which have the ability to discriminate between filamentous and pelleted growth and may have the advantage of being applicable to automated systems of culture analysis (Packer & Thomas, 1990; Patanker, Liu & Oolman, 1993). In the filamentous state growth and metabolite production proceed with exponential kinetics, but pellet growth and production follow cube-root kinetics. It is economically important to understand the transition from one form to the other. Certain parameters have been shown to have an effect upon morphology in a controllable manner, mostly through the dilution rate (Weibe & Trinci, 1991). It is the goal of these models and studies to eventually produce systems of fermentation which can utilize continuous rather than batch methods to produce maximum economic advantage (Withers *et al.*, 1994).

One important difference between submerged and surface culture is expressed as a selective difference between mycelial types (Withers *et al.*, 1994). In glucose-limited chemostat cultures, flat aconidial mutants of *Aspergillus oryzae* are favoured over wild type, a situation not found in surface cultures or natural populations. The importance of this selection to commercial production has been recognized (Weibe & Trinci, 1991). The selection of unsuitable mycelial types is exacerbated by continuous or chemostat culture. Despite the increase in cost for batch culture methods, the use of chemostat cultures for continuous production may be limited by the selection of these mycelial types.

It is unclear at this time to what extent submerged and surface morphologies are controlled by the same biochemical mechanisms. In some instances the processes known to influence surface morphologies are also known to be active in submerged culture, but whether these processes also influence submerged growth forms is unknown. Certainly, any process which relies on the diffusion of a morphogen externally will be severely effected by the large volume of diluent provided by the culture medium. The endogenous production and control of circadian rhythms continues to occur in submerged culture (Perlman, Nakashima & Feldman, 1981), which can be detected in the concentrations of the mRNAs from the circadian genes (Loros & Dunlap, 1991). This suggests that there may be other morphological processes active in liquid culture, which should allow the examination of these processes at the molecular level.

Conclusions

The fungal mycelium offers some unique opportunities for the student of development. It is a readily cultured, developmentally plastic form, which has been shown to be highly amenable in laboratory conditions. The ease with which large amounts of tissue can be grown, harvested and subjected to biochemical and molecular analysis makes mycelia excellent subjects to further a mechanistic understanding of the processes of organic development. A significant start has been made on the description of developmental events in mycelia, and the understanding of its biochemistry and developmental signals. Models which are both descriptive and provide analytical insight into developmental processes have been published and are beginning to be tested. A critical challenge to model making is to relate dynamic models of growth and differentiation to the stable, static structures seen in mycelial cultures. The processes which produce the observed patterns may not be the same as those which maintain the patterns over time. A reasonable start to building such models will probably include elements of reaction-diffusion models as well as fractal geometry and biological feedback. With the increase in sophistication of molecular techniques brought on in the last few years applied to fungi, the prospect of providing a physical basis to substantiate these models appears good. One may hope that progress in understanding the mechanisms of mycelial development will be rapid.

Acknowledgements

This work was supported in part by a grant from the graduate school of Mankato State University.

References

Anderson, M. R. & Deppe, C. S. (1976) Control of fungal development. I. The effects of two regulatory genes on growth in *Schizophyllum commune. Developmental Biology* **53**, 21–29.

Aronson, B. D., Johnson, K. A., Loros, J. J. & Dunlap, J. C. (1994). Negative feedback defining a circadian clock: autoregulation of the clock gene frequency. *Science* **263**, 1578–1584.

Arpaia, G., Loros, J. J., Dunlap, J. C., Morelli, G. & Macino, G. (1993). The interplay of light and the circadian clock. Independent dual regulation of the clock-controlled gene *ccg-2(eas)*. *Plant Physiology* **102**, 1299–1305.

Aynsley, M., Ward, A. C. & Wright, A. R. (1990). A mathematical model for the growth of mycelial fungi in submerged culture. *Biotechnology and*

Bioengineering **35**, 820–830.

Becker, P. K. & Field, R. J. (1975). Stationary patterns in the Oregonator model of the Belusov-Zhabotinskii reaction. *Journal of Physical Chemistry* **89**, 118–128.

Bourret, J. A., Lincoln, R. A. & Carpenter, B. H. (1969). Fungal endogenous rhythms expressed by spiral figures. *Science* **166**, 763–764.

Bull, A. T. & Trinci, A. P. J. (1977). The physiology and metabolic control of fungal growth. *Advances in Microbial Physiology* **15**, 1–84.

Dharmananda, S. & Feldman, J. F. (1979). Spatial distribution of circadian clock phase in aging cultures of *Neurospora crassa*. *Plant Physiology* **63**, 1049–1054.

Dowson, C. G., Rayner, A. D. M. & Boddy, L. (1986). Outgrowth patterns of mycelial cord-forming basidiomycetes from and between woody resource units in soil. *Journal of General Microbiology* **132**, 203–211.

Dowson, C. G., Springham, P., Rayner, A. D. M. & Boddy, L. (1989). Resource relationships of foraging mycelial systems of *Phanerochaete velutina* and *Hypholoma fasiculare* in soil. *New Phytologist* **111**, 501–509.

Dunlap, J. C. (1993). Genetic analysis of circadian clocks. *Annual Review of Physiology* **55**, 683–728.

Dunlap, J. C. & Feldman, J. F. (1988). On the role of protein synthesis in the circadian clock of *Neurospora crassa*. *Proceedings of the National Academy of Sciences, USA* **85**, 1096–1100.

Edelstein, L. (1982). The propagation of fungal colonies: a model for tissue growth. *Journal of Theoretical Biology* **98**, 679–701.

Edelstein, L. & Segel, L. A. (1983). Growth and metabolism in mycelial fungi. *Journal of Theoretical Biology* **104**, 187–210.

Edelstein, L., Hadar, Y., Chet, I., Henis, Y. & Segel, L. A. (1983). A model for fungal colony growth applied to *Sclerotium rolfsii*. *Journal of General Microbiology* **129**, 1873–1881.

Feldman, J. F. (1982). Genetic approaches to circadian clocks. *Annual Review of Plant Physiology* **33**, 583–608.

Fiddy, C. & Trinci, A. P. J. (1976a). Mitosis, septation, branching and the duplication cycle in *Aspergillus nidulans*. *Journal of General Microbiology* **97**, 169–184.

Fiddy, C. & Trinci, A. P. J. (1976b). Nuclei, septation, branching and growth of *Geotrichum candidum*. *Journal of General Microbiology* **97**, 185–192.

Fries, N. & Sun, Y.-P. (1992). The mating system of *Suillus bovinus*. *Mycological Research* **96**, 237–238.

Gardner, G. F. & Feldman, J. F. (1981). Temperature compensation of circadian period length in clock mutants of *Neurospora crassa*. *Plant Physiology* **68**, 1244–1248.

Gow, N. A. R. & Gooday, G. W. (1982). Growth kinetics of and morphology of colonies of the filamentous form of *Candida albicans*. *Journal of General Microbiology* **128**, 2187–2194.

Hutchinson, S. A., Sharma, P., Clarke, K. R. & MacDonald, I. (1980). Control of hyphal orientation in colonies of *Mucor hiemalis*. *Transactions of the British Mycological Society* **75**, 177–191.

Indermitte, C., Liebling, T. M. & Clemençon, H. (1994). Cultural analysis and external interaction models of mycelial growth. *Bulletin of Mathematical Biology* **56**, 633–664.

Jennings, D. H. (1987). Translocation of solutes in fungi. *Biological Reviews* **62**,

215–243.

Katz, D., Goldstein, D. & Rosenberger, R. F. (1972). Model for branch initiation in *Aspergillus nidulans* based on measurements of growth parameters. *Journal of Bacteriology* **109**, 1097–1100.

Koch, A. L. (1975). The kinetics of mycelial growth. *Journal of General Microbiology* **89**, 209–216.

Kotov, V. & Reshetnikov, S. V. (1990). A stochastic model for early mycelial growth. *Mycological Research* **94**, 577–586.

Lee, B. N. & Adams, T. H. (1994). The *Aspergillus nidulans fluG* gene is required for production of an extracellular developmental signal and is related to prokaryotic glutamine synthetase I. *Genes and Development* **8**, 641–651.

Loros, J. J., Denome, S. A. & Dunlap, J. C. (1989). Molecular cloning of genes under the control of the circadian clock in *Neurospora*. *Science* **243**, 385–388.

Loros, J. J. & Dunlap, J. C. (1991). *Neurospora crassa* clock-controlled genes are regulated at the level of transcription. *Molecular and Cellular Biology* **11**, 558–563.

Lysek, G. (1984). Physiology and ecology of rhythmic growth and sporulation in fungi. In *The Ecology and Physiology of the Fungal Mycelium* (ed. D. H. Jennings & A. D. M. Rayner), pp. 323–342. Cambridge University Press: Cambridge, U.K.

Matsuura, S. & Miyazima, S. (1993a). Colony morphology of the fungus *Aspergillus oryzae*. In *Fractals in Biology and Medicine* (ed. T. F. Nonnenmacher, G. A. Losa & E. R. Weibel). Birkhauser: Basel.

Matsuura, S. & Miyazima, S. (1993b). Colony of the fungus *Aspergillus oryzae* and self-affine fractal geometry of growth fronts. *Fractals* **1**, 1–19.

Matsuura, S. & Miyazima, S. (1993c). Formation of ramified colony of fungus *Aspergillus oryzae* on agar medium. *Fractals* **1**, 336–345.

Meinhardt, H. (1976). Morphogenesis of lines and nets. *Differentiation* **6**, 117–123.

Meinhardt, H. (1984). Models of pattern formation and their application to plant development. In *Positional Controls in Plant Development* (ed. P. W. Barlow & D. J. Carr), pp. 1–32. Cambridge University Press: Cambridge, U.K.

McLean, K. M. & Prosser, J. I. (1987). Development of vegetative mycelium during colony growth of *Neurospora crassa*. *Transactions of the British Mycological Society* **88**, 489–495.

Murray, J. D. (1993). *Mathematical Biology*, 2nd edn. Springer-Verlag: Berlin & New York.

Newell, P. C. (1983). Attraction and adhesion in the slime mold *Dictyostelium*. In *Fungal Differentiation: A Contemporary Synthesis*, Mycology Series **vol. 43** (ed. J. E. Smith), pp. 43–71. Marcel Dekker: New York.

Okon, Y., Chet, I. & Henis, Y. (1972). Lactose-induced synchronous sclerotium formation in *Sclerotium rolfsii* and its inhibition by ethanol. *Journal of General Microbiology* **71**, 465–470.

Olsvik, E., Tucker, K. G., Thomas, C. R. & Kristiansen, B. (1993). Correlation of *Aspergillus niger* broth rheological properties with biomass concentration and the shape of mycelial aggregates. *Biotechnology and Bioengineering* **42**, 1046–1052.

Packer, H. L. & Thomas, C. R. (1990). Morphological measurements on filamentous microorganisms by fully automatic image analysis. *Biotechnology and Bioengineering* **35**, 870–881.

Patankar, D. B., Liu, T. & Oolman, T. (1993). A fractal model for the characterization of mycelial morphology. *Biotechnology and Bioengineering* **42**, 571–578.

Perlman, J. , Nakashima, H. & Feldman, J. F. (1981). Assay and characteristics of circadian rhythmicity in liquid cultures of *Neurospora crassa*. *Plant Physiology* **67**, 404–407.

Piskorz-Binczycka, B. (1991). Rhythm of sporulation within the section *Penicillium clavigerum*. *Acta Biologica Cracovense* **33**, 13–25.

Prosser, J. I. (1993). Growth kinetics of mycelial colonies and aggregates of ascomycetes. *Mycological Research* **97**, 513–528.

Rayner, A. D. M. (1991). The challenge of the individualistic mycelium. *Mycologia* **83**, 48–71.

Rayner, A. D. M., Griffith, G.S. & Wildman, H.G. (1994). Induction of metabolic and morphogenetic changes during mycelial interactions among species of higher fungi. *Biochemical Society Transactions* **22**, 389–394.

Rayner, A. D. M., Powell, K. A., Thompson, W. & Jennings, D. H. (1985). Morphogenesis of vegetative organs. In *Developmental Biology of Higher Fungi* (ed. D. Moore, L. A. Casselton, D. A. Wood & J. C. Frankland), pp. 249–279. Cambridge University Press: Cambridge, U.K.

Reichl, U., King, R. & Gilles, E. D. (1992). Characterization of pellet morphology during submerged growth of *Streptomyces tendae* by image analysis. *Biotechnology and Bioengineering* **39**, 164–170.

Robinson, P. M. & Smith, J. M. (1980). Apical branch formation and cyclic development in *Geotrichum candidum*. *Transactions of the British Mycological Society* **75**, 233–238.

Sargent, M. L., Briggs, W. R. & Woodward, D. O. (1966). Circadian nature of a rhythm expressed by an invertaseless strain of *Neurospora crassa*. *Plant Physiology* **41**, 1343–1349.

Sen, R. (1990). Intraspecific variation in two species of *Sullius* from Scots pine (*Pinus sylvestris* L.) forests based on somatic incompatibility and isozyme analyses. *New Phytologist* **114**, 607–616.

Sharland, P. & Rayner, A. D. M. (1989). Mycelial ontogeny and interactions in non-outcrossing populations of *Hypoxylon*. *Mycological Research* **93**, 273–281.

Smith, M. L., Bruhn, J. N. & Anderson, J. B. (1992). The fungus *Armillaria bulbosa* is among the largest and oldest living organisms. *Nature* **356**, 428–431.

Smith, J. M. & Robinson, P. M. (1980). Development of somatic hyphae of *Geotrichum candidum*. *Transactions of the British Mycological Society* **74**, 159–165.

Steele, G. C. & Trinci, A. P. J. (1975). Morphology and growth kinetics of differentiated and undifferentiated mycelia of *Neurospora crassa*. *Journal of General Microbiology* **91**: 362–368.

Stone, J. K., Pinkerton, J. N. & Johnson, K. B. (1994). Axenic cultures of *Anisogramma anomala*: evidence for self-inhibition of ascospore germination and colony growth. *Mycologia* **86**, 674–683.

Thompson, W. & Boddy, L. (1983). Decomposition of suppressed oak trees in even-aged plantations. II. colonization of tree roots by cord- and rhizomorph-producing basidiomycetes. *New Phytologist* **93**, 277–291.

Thompson, W. & Rayner, A. D. M. (1983). Extent, development and function of mycelial cord systems in soil. *Transactions of the British Mycological Society* **81**, 333–345.

Trinci, A. P. J. (1970). Kinetics of the growth of mycelial pellets of *Aspergillus nidulans*. *Archiv für Mikrobiologie* **73**, 353–367.

Trinci, A. P. J. (1971). Influence of the peripheral growth zone on the radial growth rate of fungal colonies. *Journal of General Microbiology* **67**, 325–344.

Trinci, A. P. J. (1973a). Growth of wild type and spreading colonial mutants of *Neurospora crassa* in batch culture and on agar medium. *Archiv für Mikrobiologie* **91**, 113–126.

Trinci, A. P. J. (1973b). The hyphal growth unit of wild type and spreading colonial mutants of *Neurospora crassa*. *Archiv für Mikrobiologie* **91**, 127–136.

Trinci, A. P. J. (1974). A study of the kinetics of hyphal extension and branch initiation of fungal mycelia. *Journal of General Microbiology* **81**, 225–236.

Turing, A. (1952). The chemical basis of morphogenesis. *Philosophical Transactions of the Royal Society* **B237**, 37–72.

Viniegra-Gonzalez, G., Saucedo-Castaneda, G., Lopez-Isunza, F. & Favela-Torres, E. (1993). Symmetric branching model for the kinetics of mycelial growth. *Biotechnology and Bioengineering* **42**, 1–10.

Weibe, M. & Trinci, A. P. J. (1991). Dilution rate as a determinant of mycelial morphology in continuous culture. *Biotechnology and Bioengineering* **38**, 75–81.

Weibe, M. G., Robson, G. D., Trinci, A. P. J. & Oliver, S. G. (1992). Characterization of morphological mutants spontaneously generated in glucose-limited, continuous flow cultures of *Fusarium graminearum* A3/5. *Mycological Research* **96**, 555–562.

Withers, J. M., Weibe, M. G., Robson, G. D. & Trinci, A. P. J. (1994). Development of morphological heterogeneity in glucose-limited chemostat cultures of *Aspergillus oryzae*. *Mycological Research* **98**, 95–100.

Wittler, R., Baumgartl, H., Lubbers, D. W. & Shugerl, K. (1986). Investigations of oxygen transfer into *Penicillium chrysogenum* pellets by microprobe measurements. *Biotechnology and Bioengineering* **28**, 1024–1036.

Yamagita, T. & Kogane, F. (1962). Growth and cytochemical differentiation of mold colonies. *Journal of General and Applied Microbiology* **8**, 201–213.

Yang, H., King, R., Reichl, U. & Gilles, E. D. (1992a). Mathematical model for apical growth, septation, and branching of mycelial microorganisms. *Biotechnology and Bioengineering* **39**, 49–58.

Yang, H., Reichl, U., King, R. & Gilles, E. D. (1992b). Measurement and simulation of the morphological development of filamentous organisms. *Biotechnology and Bioengineering* **39**, 44–48.

Chapter 4

The genetics of morphogenesis in *Neurospora crassa*

P. J. VIERULA

Summary

The morphology of fungi is thought to arise from the epigenetic interactions of myriad genes involved in growth and the response of these genes to environmental stimuli. The extensive genetic investigations, which have been undertaken over the years with *N. crassa*, lend credence to this hypothesis. Approximately 25% of the more than 500 mutants described to date have altered morphologies. While some of these strains have genetic lesions which specifically affect conidiation or development of sexual reproductive structures, a number of strains have defects in vegetative morphogenesis. In some cases, the nature of the lesion has been inferred from cytological observations or the enzyme activities in cell extracts. More recently, molecular biology has begun contributing to our understanding of genes which affect morphogenesis. This chapter reviews from a primarily genetic standpoint: morphological programs in *Neurospora*, morphological mutants, cell wall formation, septation and branching, the cytoskeleton and apical growth, the influence of the mitochondrion, environmental influences, pleiotropic effects, and lessons which can be derived from yeast.

Introduction

An understanding of the formative processes which shape an organism continues to be an important challenge for biologists. While most of the attention has focused on complex multicellular organisms, the microbial world offers many fascinating examples of distinctive and elaborate morphologies which are well suited to experimental manipulation (Harold, 1990).

One of these micro-organisms, the fungus *Neurospora crassa*, has a long history as an experimental subject for studies of morphogenesis and development (Scott, 1976; Mishra, 1977). Its rapid growth rate, tractable genetics and more recently, its amenity to molecular genetic manipulation continue to make it an attractive model organism. The branched, tubular hyphae of *Neurospora* are formed by apical growth, the periodic initiation of new branches from established hyphae and interhyphal fusions. Hyphal compartments are haploid and multinucleate, and adjacent compartments are separated by perforated cross walls which allow the movement of cytoplasm and organelles. Distinctive reproductive structures superimposed on this primary vegetative thallus give rise to the mature colony.

In response to desiccation or carbon limitation, some hyphae turn and grow away from the substrate to initiate asexual sporulation (reviewed by Springer, 1993). These aerial hyphae produce conidiophores which give rise to multinucleate macroconidia by a process of repeated apical budding. Limited numbers of small uninucleate microconidia are also formed by extrusion through a rupture in the walls of vegetative hyphae. The sexual cycle is initiated by nitrogen limitation which induces formation of compact, coiled hyphae known as the female reproductive structures (protoperithecia). These differentiated structures produce thin receptive hyphae, the trichogynes, which grow towards and ultimately fuse with cells of the opposite mating type. Following fertilization, protoperithecia enlarge and darken to become the perithecia within which the meiotic spores develop.

A genetic approach to morphogenesis

Genetic studies of morphogenesis in *N. crassa* were initiated by E. L. Tatum and his colleagues with the isolation of a large collection of morphological mutants (Garnjobst & Tatum, 1967). These were assembled into 6 broad categories: (i) true colonials, (ii) spreading colonials, (iii) semi-colonials, (iv) spreading morphologicals, (v) distinctive morphologicals and (vi) morphologicals which are dependent on particular environmental conditions to be manifest. Examples of some of these different colony morphologies, compared to wild type, are shown in Figs 1 to 6. Considerable variation in colony diameter is evident in the true colonial mutants such as *doily* (Fig. 1), *peak* (Fig. 2), and *col-5* (Fig. 3). The distinctive colony of *ro-1* (Fig. 4) results from the braiding of parallel hyphae. The colonies of spreading colonial forms such *spco-4* (Fig. 5) have a slower growth rate but resemble the wild type (Fig. 6). Many new mutants, in each of the above categories continue to be added to the collection (Perkins *et al.*, 1982).

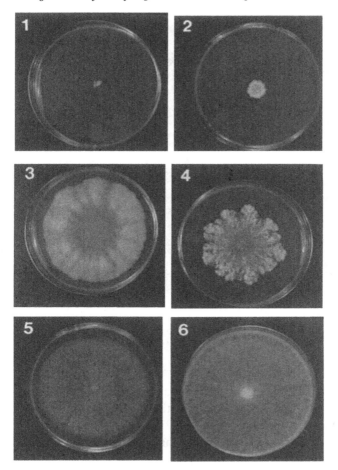

Figs 1-6. Photographs of colonies grown on minimal medium for 60 hours at 23°C (Fig. 1-4) or for 48 hours at 23°C. *N. crassa* mutants *doily* (Fig. 1), *peak* (Fig. 2), *col*-5 (Fig. 3), *ro*-1 (Fig. 4), *spco*-4 (Fig. 5) and wild type (Fig. 6).

Many morphological mutants exhibit an increase in the number of branches per unit length of hypha. Examples of this can be seen with *ro*-1 (Fig. 9) and *peak* (Fig. 10) compared to wild-type (Fig. 7). It seems intuitive that any impairment in the rate of hyphal extension, which does not also affect branch initiation, would result in an apparent increase in branching frequency. However, some strains such as *col*-5 (Fig. 8) exhibit little if any change in branching frequency although hyphal outgrowth is clearly

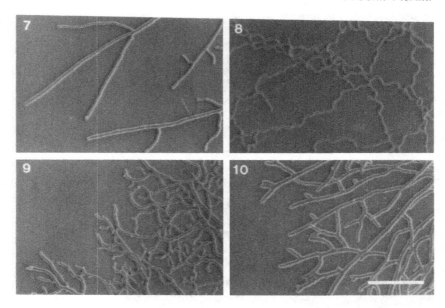

Figs 7-10. Low magnification scanning electron micrographs of hyphae of wild-type (Fig. 7), *col*-5 (Fig. 8), *ro*-1 (Fig. 9), and *peak* (Fig. 10). Bar = 100 μm.

anomalous. Therefore, hyphal extension and branch initiation are likely to be relatively independent processes. Studies with other fungi also support this hypothesis (Wiebe, Robson & Trinci, 1992).

Morphological mutants often form dense compact colonies which are difficult to distinguish with the naked eye. However, microscopic examination of these colonies can reveal very striking differences in morphology compared to wild-type (Fig. 11). Growth of the *doily* strain (Fig. 12) is still polarized and therefore filamentous, but the hyphal extension rate is markedly reduced. In comparison, both of the colonial temperature-sensitive strains *cot*-2 (Fig. 13) and *cot*-1 (Fig. 14) appear to have defects in polarized growth when incubated at a restrictive temperature.

In an effort to assign a likely biochemical defect to each mutation, many of the strains have been subjected to extensive biochemical analyses (Scott, 1976; Mishra, 1977). For example, a study of hyphal cell wall structure of 11 colonial or semicolonial mutants found that the levels of one or more structural polysaccharides is altered in all of the single-gene mutants (Mahadevan & Tatum, 1965). Similarly, an analysis of the cell wall composition of 23 different morphological mutants revealed that all of

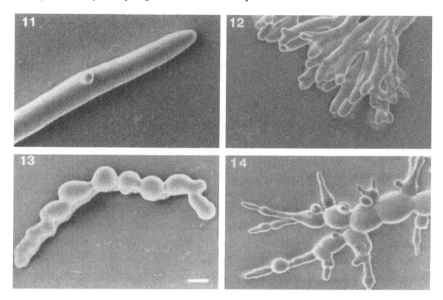

Figs 11-14. High magnification scanning electron micrographs of hyphae of wild-type (Fig. 11), *doily* (Fig. 12), *cot*-2 grown at 37C (Fig. 13) and *cot*-1 germinated at 23°C and shifted to 37°C for 6 hours (Fig. 14). Bar = 10 μm.

them had reductions in hyphal wall peptides (Wrathal & Tatum, 1974). Since these changes cannot be causally related to all of the morphological mutations examined, it appears likely that lesions which impair hyphal extension have pleiotropic effects on cell wall composition. Therefore, correlations between morphology and biochemical alterations need to be interpreted with caution.

Of the more than 100 mutants with abnormal vegetative phenotype which have now been described, many also exhibit some changes in both conidiogenesis and development of sexual reproductive structures (Perkins *et al.*, 1982). This is not surprising since both developmental programmes are also dependent on the fundamental process of apical growth. Despite the overlap, both conidiation and formation of protoperithecia involve the expression of unique subsets of genes not involved in growth of the vegetative mycelium (Berlin & Yanofsky, 1985; Nelson & Metzenberg, 1992; Springer, 1993). Therefore, it is possible that the aberrant vegetative morphology of some of the mutants results from mis-scheduled expression of developmentally regulated genes.

In this chapter I will focus primarily on the genetics of morphogenesis as it pertains to the growth of the vegetative thallus. A comprehensive review of developmental pathways in *Neurospora* has appeared recently (Springer, 1993).

The role of the cell wall

A cogent demonstration of the importance of the cell wall in determining fungal morphology is provided by the enzymatic removal of the cell wall from hyphal filaments. The resulting protoplasts are spherical in shape and must be protected against osmolysis. Hyphal growth resumes only if the wall is allowed to regenerate.

The laminated *Neurospora* cell wall has an inner layer enriched in chitin microfibrils which is believed to be overlaid by a secondary layer of proteins and then an amorphous matrix of glycoproteins and glucans (Hunsley & Burnett, 1970). Both chitin and glucans are thought to be responsible for the inherent structural rigidity of the fungal wall. Consequently, all of the biochemical processes involved in precursor biosynthesis, transport of precursors and enzymes to the growing tip and finally cell-wall synthesis are potential targets for mutational disruption. This has been borne out by several recent studies which have focused on the biosynthesis of two principal cell-wall polymers, chitin and glucan, for genetic analysis.

The chitin synthase gene of *N. crassa* was cloned by the polymerase chain reaction (PCR) using primers based on the *CHS*1 and *CHS*2 genes which encode this enzyme in *S. cerevisiae* (Yarden & Yanofsky, 1991). The predicted amino acid sequence encoded by the *Neurospora* gene is 37% and 39% identical to the two yeast gene products and it has several potential membrane-spanning domains near the carboxyl terminus as expected for a membrane-associated enzyme. The *Neurospora chs-*1 gene was inactivated by repeat induced point mutation (RIP) with approximately 40% of progeny exhibiting a slow growth phenotype. One of these new mutant strains created by RIP, *chs-*1RIP, grows at less than 15% of the wild type rate as assessed by measuring colony diameter. In addition, it has a lower hyphal density, and conidiation is sparse. Swelling of both apical, and sub-apical hyphal compartments was evident and the tips of thin aerial hyphae were often deteriorated. Inactivation of *chs-*1 had no apparent effect on septation and frequency of branching. Chitin synthase activity in this strain was 7- to 20-fold lower than wild-type with residual activity attributed to other chitin synthases in the cell.

As expected, a second chitin synthase gene, *chs-*2, was also discovered (Din & Yarden, 1994). In this case, however, gene inactivation experiments using RIP had no demonstrable effect on morphology or development.

Despite reduced chitin synthase activity and an increased sensitivity to edifenphos, an inhibitor of phosphatidylcholine biosynthesis, the *chs*-2RIP mutant strain had normal cell-wall chitin content. Din & Yarden (1994) speculate that this enzyme could have a specialized role in repair. Presumably any deficiency in *CHS*-2 activity is compensated for by *CHS*-1 and perhaps other chitin synthases.

The other major polymer in *Neurospora* cell walls is amorphous β-glucans in β-1,3- and β-1,6-linkages (Mahadevan & Tatum, 1965). One remarkable strain known as 'slime' has no cell-wall and grows as a plasmodium in hypertonic medium (Emerson, 1963). The slime phenotype requires the presence of three genes simultaneously, *fuzzy* (*fz*), *spontaneous germination* (*sg*) and *osmotic* (*os*) (Emerson, 1963; Leal-Morales & Ruiz-Herrera, 1985). Only the *os* mutation is known to affect cell walls; the contribution of the other mutations is unknown.

One of the approaches used to study glucan biosynthesis is to generate 'wall-less' strains, such as the *os*-1 strain which is deficient in (1,3)β-glucan synthase activity, by ethylmethanesulfonate mutagenesis (Phelps *et al.*, 1990). A total of 22 mutants, all with significantly reduced β-glucan synthase activity, were isolated. One of these strains was subsequently used to clone a gene by complementation, designated glucan synthase 1 (*gs*-1) gene (Enderlin & Selitrennikoff, 1994). Although this gene restores 30 to 40% of glucan synthase activity to the mutant strain, the deduced 59 kDa protein encoded by *gs*-1 has no transmembrane domains or signal peptide cleavage sites and it is therefore unlikely to be a biosynthetic enzyme. Instead, it has a limited similarity to a yeast transcriptional regulatory protein Knr4/Smi1, also implicated in (1,3)β-glucan synthesis. Therefore, the *gs*-1 gene product has been proposed to be involved in the regulation of glucan biosynthesis, perhaps functioning as a DNA binding protein (Enderlin & Selitrennikoff, 1994).

A new mutant strain was constructed in *os*-1 by replacement of the *gs*-1 gene with a selectable drug resistance marker. This new *gs*-1D7, *os*-1 strain has the same low levels of (1,3)-β-glucan synthase as the original mutant. The effect of deletion of the *gs*-1 gene on enzyme levels and morphology in a wild-type genetic background has not yet been reported, but it is likely that phenotypic expression is dependent on the presence of the *os*-1 lesion. Analysis of additional members of the glucan synthase-deficient collection of mutants may help clarify the actual role of the *gs*-1 gene product.

One mutant, named strain 18, with an altered composition of cell wall β-glucans has also been reported (Chiba, Nakajima & Matsuda, 1988). The number of 1,3-linkages in the alkali-soluble fraction of this mutant was dramatically reduced, and the mutant glucan had 2.5 times as many 1,6 branches. The mutant is reported to grow more slowly than wild-type, and cellular compartments ·are more rounded, giving hyphae a beaded morphology in mature cultures (Chiba *et al.*, 1988). No data is available on the relationship between the mutation in this strain and other mutants with altered cell-wall composition such as *os-1*.

Primary metabolism and morphology

Before the era of recombinant DNA technology, a considerable effort was devoted to attempts to identify significant changes in some enzyme activities which could be attributed to each morphological mutation. The results of this work are summarized in Table 1, and readers are referred to several earlier reviews for a more comprehensive account (Brody, 1973; Scott, 1976; Mishra, 1977).

One of the significant enzyme deficiencies identified in the *ragged* (*rg*) mutants is phosphoglucomutase (PGM) (Brody & Tatum, 1966). This activity catalyses the interconversion between glucose-1-phosphate (G-1-P) and glucose-6-phosphate (G-6-P), an important branch point in carbohydrate metabolism. The two *rg* loci are thought to encode two isozymes which together make up the active PGM enzyme. The decrease in cell-wall β-1,3-glucan and ragged morphology may be a direct result of G-1-P depletion or allosteric inhibition of other enzymes by excessive accumulation of G-6-P (Brody & Tatum, 1967; Mishra, 1977).

The reduced levels of glucose-6-phosphate dehydrogenase (G6PD) reported for mutants *col*-2, *balloon* and *frost* have been proposed to alter morphology indirectly because loss of G6PD activity would lead to a depletion of NADPH (Scott & Tatum, 1970; Scott, 1976). Since NADPH is required for fatty acid biosynthesis, the morphological change was proposed to be caused by the alterations in cell membrane (Brody, 1973). This hypothesis is supported by the fact that choline- and inositol-deficient mutants have similar morphological abnormalities when deprived of these membrane constituents (Mishra, 1977).

Another enzyme from the pentose phosphate pathway, 6-phosphogluconate dehydrogenase (6PGD) reportedly has altered kinetics in two morphological strains *col*-3 and *col*-10 (Scott & Abramsky, 1973). Surprisingly, 6PGD activity is normal in both mutant strains. It is unclear how the altered property, a 3-fold increase in K_m, could have such a profound effect on morphology.

Table 1. *Morphological mutants of* N. crassa *to which a gene function has been assigned*

Mutant	Phenotype	Function	Reference
Primary metabolism			
col-2	colonial	glucose-6-phosphate dehydrogenase	Brody & Tatum, 1966
bal	colonial	glucose-6-phosphate dehydrogenase	Brody & Tatum, 1966
frost	spreading colonial	glucose-6-phosphate dehydrogenase	Brody & Tatum, 1966
rg-1	colonial	phosphoglucomutase	Brody & Tatum, 1967
rg-2	colonial	phosphoglucomutase	Brody & Tatum, 1967
col-3	colonial	6-phosphogluconate dehydrogenase	Brody & Tatum, 1966
col-10	colonial	6-phosphogluconate dehydrogenase	Brody, 1973
inos	colonial (without inositol)	inositol phosphate synthase	Brody, 1973
Cell wall synthesis			
chs-1[RIP]*	abnormal	chitin synthase 1	Yarden & Yanofsky, 1991
gs-1::Hyg*	osmotic	regulatory protein	Enderlin & Selitrennikoff, 1994
gls-1[18]	abnormal	glucan synthase (1,3-linkage)?	Chiba et al., 1988
Cytoskeleton			
*ro-1**	ropy	cytoplasmic dynein	Plamann et al., 1994
*ro-2**	ropy	unknown	Chiba et al., 1988; Vierula & Mais (unpublished)
*ro-3**	ropy	glued/dynactin	Plamann et al., 1994
*ro-4**	ropy	actin-related protein	Plamann et al., 1994; Robb et al., 1995
sn	snowflake	involved in filament assembly	Alvarez et al., 1993
Signalling and regulation			
*cot-1**	colonial (ts)	protein kinase C	Yarden et al., 1992
*cr-1**	restricted	adenylate cyclase (*nac*)?	Kor-eda et al., 1991

*indicates that the gene has been cloned.

The mitochondrion and morphogenesis

As *Neurospora* is an obligate aerobe, mutations in nuclear genes encoding components of the electron transport chain result in a predictable slow-growth phenotype (Mitchell, Mitchell & Tissieres, 1953; Harkness *et al.*, 1995). Some extranuclear mutants such as *poky*, which has a 4 bp deletion in the large mitochondrial rRNA gene also exhibit slow continuous growth (Mitchell & Mitchell, 1952; Akins & Lambowitz, 1984). Both resemble leaky auxotrophic mutants which have defects in essential genes involved in primary metabolism.

However, another group of mutants, with deletions in their mitochondrial genomes, have a novel 'stop-start' phenotype in which the hyphae alternately grow and stop for an indeterminate period of time varying from a few hours to several days (Bertrand *et al.*, 1980). As with *poky* mutants, these strains are heteroplasmons, containing both a defective mitochondrial genome and low levels of wild-type or less damaged genomes which permit growth.

It has been proposed that the defective mitochondrial genomes have some replicative advantage which results in their preferential accumulation at the mycelial front. Growth stoppage at the front results from the subsequent death of the hyphal tips attributable to a loss of mitochondrial function below a certain minimum threshold. Growth resumes from hyphal compartments distal to the mycelial front which still have a viable ratio of functional to non-functional mitochondrial genomes (Bertrand *et al.*, 1980). The suppressive accumulation of the defective mitochondrial genome then gradually kills hyphal tip cells by the attendant loss of oxidative phosphorylation. Some of these stopper mutants have arisen by the integration of variant mitochondrial plasmids into the organellar genome, and subsequent accompanying deletions and rearrangements (Akins, Kelley & Lambowitz, 1986).

Signalling and regulatory circuits

The *crisp*-1 (*cr*-1) mutant has a restricted colony morphology and an earlier and more exuberant conidiation compared to wild-type (Terenzi, Fláwia & Torres, 1974). The *cr*-1 locus has been proposed to encode an adenylate cyclase since the specific activity of this enzyme is significantly lower in the mutant strain. Exogenous cAMP added to the growth medium can overcome the pleiotropic effects of this mutation (Terenzi *et al.*, 1976).

The decrease in adenylate cyclase activity is not unique to the *cr*-1 strains.

The *frost* mutant also has a reduced specific activity but the severity of the morphological changes can be alleviated by supplementation of the growth medium with linolenic acid (Scott & Solomon, 1974). Therefore, it is possible that the change in adenyl cyclase activity in *cr*-1 results from some impairment in membrane function (Mishra, 1977).

An adenylate cyclase gene (*nac*) has been cloned from a *Neurospora* genomic library using the adenylate cyclase gene from *Saccharomyces cerevisiae* as a probe (Kor-eda, Murayama & Uno, 1992). The predicted polypeptide is 2300 amino acids in length and it is 28% identical to the yeast enzyme. The cloned *nac* DNA was mapped to the right arm of chromosome I, the same general region known to contain the *cr*-1 locus suggesting that *cr*-1 may be an allele of the *nac* gene (Kor-eda *et al.*, 1992). With the clone now available, it should be possible to confirm this by complementation studies and generate new mutant alleles of the *nac* gene by disruption with the RIP technique or by gene replacement.

The *cot*-1 gene appears to play an important role in hyphal morphogenesis, since mutations have a very profound effect on morphology. At temperatures greater than 32°C, hyphae of the *cot*-1 strain form many intercalary septa and branch prodigiously (Steele & Trinci, 1977; Collinge, Fletcher & Trinci, 1978). Branches which form at the elevated temperature tend to be thin and tapered and many of the tips show signs of deterioration (Fig. 3). In contrast, hyphal compartments which have formed before a temperature upshift increase in diameter and become rounded. During prolonged incubation at elevated temperatures, the wall thickness of the *cot*-1 mutant doubles (Collinge *et al.*, 1978).

The *cot*-1 gene encodes a polypeptide related to members of the cAMP-dependent protein kinase family (Yarden *et al.*, 1992). In addition to the conserved catalytic domain, the predicted COT1 protein has a long amino terminal domain which may be important for determining substrate specificity. When wild type spheroplasts were transformed with a disrupted construct, over 90% of primary transformants had a severe colonial defect. This result was surprising since only the first 209/622 amino acids of the *cot*-1 gene in the construct could be expressed, and the majority of primary transformants would be expected to be heterokaryons. Yarden *et al.* (1992) suggest that either (i) the ectopically expressed, truncated COT1 polypeptide acts in *trans* to interfere with the wild type *cot*-1 gene product or, (ii) that the disrupted construct titrates some factor required for expression of the wild type gene. No proteins have yet been identified which might be regulated by the phosphorylating activity of COT1.

The *snowflake* (*sn*) mutant of *N. crassa* has a dramatically reduced rate of hyphal extension and increased branching per unit length (Temporini & Rosa, 1993). Two alleles of *sn* have dramatically different effects on asexual reproductive structures. While strain *sn*C136 produces aerial hyphae and enlarged, spherical blastoconidiospores, *sn*JL301 has truncated aerial hyphae and yields only arthrospores. Experiments creating forced hetero-karyons have demonstrated that *sn*JL301 is recessive to wild-type and *sn*C136 is semi-dominant. Both *sn* strains are missing a 48 kDa protein named P59Nc of unknown function.

Although the *sn* locus has not yet been characterized, some clues about a possible function have emerged from a recent characterization of the protein P59Nc. The *sn* strains accumulate abnormal sized and shaped cytoplasmic and nuclear bundles of P59Nc, a protein which assembles into 8 to 10 nm filaments in both mutant and wild-type hyphae (Rosa, Alvarez & Maldonado, 1990). The gene encoding P59Nc, *cfp* has been cloned and shown to encode a pyruvate decarboxylase (PDC) (Alvarez *et al.*, 1993). When *N. crassa* hyphae are grown on ethanol medium and consequently depleted of PDC/P59Nc, hyphal morphology is not dramatically altered (Alvarez *et al.*, 1993). Therefore the PDC/P59Nc filaments appear to have no direct role in morphogenesis. However, the *sn* gene product may influence morphology by facilitating the assembly of cytoskeletal proteins (Haedo *et al.*, 1992).

Role of the cytoskeleton

Unlike animal cells, hyphal shape is maintained by the external wall, not by an internal network of structural girders. An exception to this may be at the growing hyphal tip where the expanding cell wall is thin and malleable (Barja, Nguyen Thi & Turian, 1991). The dense network of actin filaments may anchor the plasma membrane internally during apical cell-wall expansion.

The cytoskeleton also plays other important roles within the cell such as the localization of macromolecules, and the movement and positioning of organelles. One group of morphological mutants which have proved to be particularly valuable for the genetic characterization of the cytoskeleton is *ropy* (*ro*). Hyphae of these strains are much more curled and branched than wild-type, with cable-like intertwining of aerial hyphae giving the *ro* colonies a characteristic radial pattern.

Cytological examination of the *ro* mutants has also revealed a defect in nuclear migration. In wild-type hyphae, nuclei are relatively evenly spaced and the apical nucleus moves in concert with the growing tip. In contrast, nuclei in *ro* strains are found in anomalous clusters and apical compartments are frequently devoid of nuclei although mitochondrial distribution appears to be normal. To date, molecular cloning and characterization of several of the *ro* genes has shown that they encode cytoskeletal components. The dynein heavy chain, coded for by the *ro-1* gene, is a 495 kDa subunit of a mechanochemical enzyme which mediates minus-end-directed movement along microtubules (Plamann *et al.*, 1994). One of the surprises to emerge from studies on this group of mutants and from related studies with yeasts and vertebrate cells was the discovery that an actin-related protein, encoded by the *ro-4* gene, is also involved in this microtubule-based motility system (Plamann *et al.*, 1994; Robb, Wilson & Vierula, 1995). It has been proposed that the actin-related proteins may serve to link the network of actin filaments to the microtubule system (Muhua, Karpova & Cooper, 1994). For filamentous fungi such as *Neurospora*, the apical nucleus may be tethered to the hyphal tip by a linkage between cytoplasmic microtubules nucleated by the spindle pole body and the dense network of actin microfilaments at the hyphal tip.

The other *ro* loci may encode other components of the multicomponent dynein motor complex. Alternatively, since nuclear movements are a carefully regulated process, they may also encode signalling or regulatory systems. The *ro-2* gene encodes a novel protein with high similarity to known regulatory proteins (Vierula & Mais, unpublished). All 8 of the original *ro* mutations have been shown to be partial suppressors of the *cot-1* protein kinase. Since COT1 is unlikely to directly regulate all *ro* gene products, Plamann *et al.* (1994) have proposed that the motility defect itself leads to a partial over-ride of the *cot-1* regulatory system.

The relationship between the nuclear migration defect and the aberrant morphology of the *ro* strains is still open to conjecture. It is possible that the motility defect has less obvious effects on other processes such as vesicle transport to the growing tip. Alternatively, the excessive curling and branching of *ro* hyphae suggests that the apical nucleus plays a role in directing polarized growth. Disruption of these interactions would then lead to a more random movement of the growing hyphal tip.

Environmental influences

Morphogenesis can also be very responsive to environmental conditions such as nutrient availability which can significantly influence both the vegetative thallus and development of reproductive structures. For example, when ethanol or acetate is provided as the sole carbon source instead of sucrose, the mycelial front advances at the same rate but growth proximal to the front is reduced. Leaky mutations affecting primary metabolism often result in a similar sparse growth phenotype.

The macroscopic growth patterns of filamentous fungi are also very responsive to a host of chemicals. In some cases, these changes in morphology can be quite dramatic. For example, early studies by Tatum, Barratt & Cutter (1949) identified a number of paramorphogens including sorbose, gammexane and non-ionic detergents such as Tergitol which all induce colonial growth.

If the cellular mechanisms of these chemicals are known, a comparison of morphologies created by mutation and chemical induction can provide useful corroborative evidence about the genetic lesion. For example, the morphology of cultures treated with gammexane, an inhibitor of inositol metabolism, is similar to those of *inositol* (*inos*) mutants (Tatum *et al.*, 1949).

Experiments which examine the effects of some paramorphogens also have a predictive value. One example of this might be the colonial growth induced by incorporation of lithium into the medium (Hanson, 1991). Assays of inositol 1-phosphate and inositol 4-phosphate phosphatases showed that both enzymes were strongly inhibited by a relatively wide range of lithium concentrations (Hansen, 1991). It is reasonable to expect that genetic lesions in the genes encoding these enzymes will have comparable effects on morphology. However, lithium has an inhibitory effect on multiple enzymes and mutations in a single gene may not sufficient for expression of phenotype. The wall-less phenotype of the slime strain described earlier provides a very apt illustration of this potential problem.

Lessons from yeast

Cellular morphogenesis of yeast cells is closely coordinated with cell cycle events (Chenevert, 1994). In the fission yeast *Schizosaccharomyces pombe* the most common morphological mutants have a rounded phenotype indicative of a loss of growth polarization (Madden, Costigan & Snyder, 1992). Similarly, one group of *Saccharomyces cerevisiae* morphological mutants, exemplified by *cdc*24 and *cdc*42, fail entirely to produce buds and grow at the restrictive temperature as large, rounded multinucleate cells (Pringle & Hartwell, 1981). These strains no longer exhibit polarized growth, and new cell wall deposition occurs randomly. Other classes of mutants produce abnormal elongate buds that fail to divide. In some of these abnormal budding strains such as *cdc*4, the nuclear cycle arrests, while in others, such as *cdc*3, nuclei continue to divide giving rise to multinucleate cells. A fourth class, comprised of the *bud* mutants, produce morphologically normal buds but bud-site selection is altered (Chant, 1994).

Many of the yeast genes have now been cloned and characterized. Some, such as *cdc3* encode filament proteins which localize to the bud neck and are presumed to play a structural role. The gene products of *cdc24* and *cdc42* resemble the *ras*-like regulatory proteins. The *bud* gene products resemble components of signal transduction pathways, nucleotide exchange factors or regulatory molecules. One of the principal themes to emerge from these extensive studies with yeasts is that polarized growth involves a large number of both structural and regulatory proteins which function in a precise temporal sequence.

We should anticipate finding *Neurospora* homologues to most if not all of these yeast proteins. For example, proteins like *cdc24* and *cdc42* may be essential for tip growth and branch initiation. Defects in *bud*-like genes may have an aberrant branching pattern although apical extension should be largely unaffected. Some of these genes may be discovered by complementation of existing mutants or employing a more direct approach using heterologous hybridization or PCR.

Conclusions

Harold (1990) has argued that, unlike metazoan organisms, microbial morphology is largely an epigenetic phenomenon and not specified directly by a genetic programme. If it is true, the vegetative morphology of fungi would be in essence the manifestation of the concerted actions and interactions of a diverse array of gene products required for vegetative growth. The studies with *Neurospora* described here are generally supportive of this view since the limited number of genes characterized to date all appear to play some important role in growth. The effect on morphology appears to be a secondary consequence of the genetic lesion.

Neurospora is a useful system for studying the complex processes of polarized growth because both dramatic and subtle genetic changes can be readily detected in the altered macroscopic growth patterns of the colony. The extensive existing collections of morphological mutants are one testament to the potential utility of this organism.

References

Akins, R. A., Kelley, R. L. & Lambowitz, A. M. (1986). Mitochondrial plasmids of *Neurospora*: integration into mitochondrial DNA and evidence for reverse transcription in mitochondria. *Cell* **47**, 505–516.

Akins, R. A. & Lambowitz, A. M. (1984). The [*poky*] mutant of *Neurospora* contains a 4-base-pair deletion at the 5' end of the mitochondrial small rRNA. *Proceedings of the National Academy of Sciences, U.S.A.* **81**, 3791–3795.

Alvarez, M. E., Rosa, A. L., Temporini, E. D., Wolstenholme, A., Panzetta, G., Patrito, L. & Maccioni, H. J. F. (1993). The 59-kDa polypeptide constituent of 8–10 nm cytoplasmic filaments in *Neurospora crassa* is a pyruvate decarboxylase. *Gene* **130**, 253–258.

Barja, F., Nguyen Thi, B.-N. & Turian, G. (1991). Localization of actin and characterization of its isoforms in the hyphae of *Neurospora crassa*. *FEMS Microbiology Letters* **77**, 19–24.

Berlin, V. & Yanofsky, C. (1985). Protein changes during the asexual cycle of *Neurospora crassa*. *Molecular and Cellular Biology* **5**, 839–848.

Bertrand, H., Collins, R. A., Stohl, L. L., Goewert, R. R. & Lambowitz, A. M. (1980). Deletion mutants of *Neurospora crassa* mitochondrial DNA and their relationship to the 'stop-start' growth phenotype. *Proceedings of the National Academy of Sciences, U.S.A.* **77**, 6032–6036.

Brody, S. (1973). Metabolism, cell walls and morphogenesis. In *Developmental Regulation: Aspects of Cell Differentiation* (ed. S. J. Coward), pp. 107–154. Academic Press: New York.

Brody, S. & Tatum, E. L. (1966). The primary biochemical effect of a morphological mutation in *Neurospora crassa*. *Proceedings of the National Academy of Sciences, U.S.A.* **56**, 1290–1297.

Brody, S. & Tatum, E. L. (1967). Phosphoglucomutase mutants and morphological changes in *Neurospora crassa*. *Proceedings of the National Academy of Sciences, U.S.A.* **58**, 423–430.

Chant, J. (1994). Cell polarity in yeast. *Trends in Genetics* **10**, 328–333.

Chenevert, J. (1994). Cell polarization directed by extracellular cues in yeast. *Molecular Biology of the Cell* **5**, 1169–1175.

Chiba, Y., Nakajima, T. & Matsuda, K. (1988). A morphological mutant of *Neurospora crassa* with defects in the cell wall β-glucan structure. *Agricultural and Biological Chemistry* **52**, 3105–3111.

Collinge, A. J., Fletcher, M. H. & Trinci, A. P. J. (1978). Physiology and cytology of septation and branching in a temperature-sensitive colonial mutant (*cot* 1) of *Neurospora crassa*. *Transactions of the British Mycological Society* **71**, 107–120.

Din, A. B. & Yarden, O. (1994) The *Neurospora crassa* chs-2 gene encodes a non-essential chitin synthase. *Microbiology* **140**, 2189–2197.

Emerson, S. (1963). Slime, a plasmodioid variant of *Neurospora crassa*. *Genetica* **34**, 162–182.

Enderlin, C. S. & Selitrennikoff, C. P. (1994). Cloning and characterization of a *Neurospora crassa* gene required for (1,3)β-glucan synthase activity and cell wall formation. *Proceedings of the National Academy of Sciences, U.S.A.* **91**, 9500–9504.

Garnjobst, L. & Tatum, E. L. (1967). A survey of new morphological mutants in *Neurospora crassa*. *Genetics* **57**, 579–604.

Haedo, S. D., Temporini, E. D., Alvarez, M. E., Maccioni, H. J. F. & Rosa, A. L. (1992). Molecular cloning of a gene (*cfp*) encoding the cytoplasmic filament protein P59Nc and its genetic relationship to the *snowflake* locus of *Neurospora crassa*. *Genetics* **131**, 575–580.

Hanson, B. A. (1991) The effects of lithium on the phosphoinositides and inositol phosphates of *Neurospora crassa. Experimental Mycology* **15**, 76–90.

Harkness, T. A. A., Rothery, R. A., Weiner, J. H., Werner, S., Azevedo, J. E., Videira, A. & Nargang, F. E. (1995). Disruption of the gene encoding the 78-kilodalton subunit of the peripheral arm of complex I in *Neurospora crassa. Current Genetics* **27**, 339–350.

Harold, F. M. (1990). To shape a cell: an inquiry into the causes of morphogenesis of microorganisms. *Microbiological Reviews* **54**, 381–431.

Hunsley, D. & Burnett, J. H. (1970). The ultrastructural architecture of the walls of some hyphal fungi. *Journal of General Microbiology* **62**, 203–218.

Kor-eda, S., Murayama, T. & Uno, I. (1991). Isolation and characterization of the adenylate cyclase structural gene of *Neurospora crassa. Japanese Journal of Genetics* **66**, 317–334.

Leal-Morales, C.A. & Ruiz-Herrera, J. (1985). Alterations in the biosynthesis of chitin and glucan in the slime mutant of *Neurospora crassa. Experimental Mycology* **9**, 28–38.

Madden, K., Costigan, C., & Snyder, M. (1992). Cell polarity and morphogenesis in *Saccharomyces cerevisiae. Trends in Cell Biology* **2**, 22–29.

Mahhadevan, P. R. & Tatum, E. L. (1965). Relationship of the major constituents of the *Neurospora crassa* cell wall to wild-type and colonial morphology. *Journal of Bacteriology* **90**, 1073–1081.

Mishra, N. C. (1977). Genetics and biochemistry of morphogenesis in *Neurospora. Advances in Genetics* **19**, 341–405.

Mitchell, M. B. & Mitchell, H. K. (1952). A case of 'maternal' inheritance in *Neurospora crassa. Proceedings of the National Academy of Sciences, U.S.A.* **38**, 442–449.

Mitchell, M. B., Mitchell, H. D. & Tissieres, A. (1953). Mendelian and non-mendelian factors affecting the cytochrome system in *Neurospora crassa. Proceedings of the National Academy of Sciences, U.S.A.* **39**, 606–613.

Muhua, L., Karpova, T. S. & Cooper, J. A. (1994) A yeast actin-related protein homologous to that in vertebrate dynactin complex is important for spindle orientation and nuclear migration. *Cell* **78**, 669–679.

Nelson, M. A. & Metzenberg, R. L. (1992). Sexual development genes of *Neurospora crassa. Genetics* **132**, 149–162.

Perkins, D. D., Radford, A., Newmeyer, D. & Bjrkman, M. (1982). Chromosomal loci of *Neurospora crassa. Microbiological Reviews* **46**, 426–570.

Phelps, P., Stark, T. & Selitrennikoff, C. P. (1990). Cell wall assembly of *Neurospora crassa*: isolation and analysis of cell wall-less mutants. *Current Microbiology* **21**, 233–242.

Plamann M., Minke, P. F., Tinsley, J. H. & Bruno, K. S. (1994). Cytoplasmic dynein and actin-related protein Arp1 are required for normal nuclear distribution in filamentous fungi. *Journal of Cell Biology* **127**, 139–149.

Pringle, J. R. & Hartwell, L. H. (1981). The *Saccharomyces cerevisiae* cell cycle. In *The Molecular Biology of the Yeast* Saccharomyces: Life Cycle and Inheritance (ed. J. N. Strathern, E. W. Jones & J. R. Broach), pp. 97–142. Cold Spring Harbor Laboratory: Cold Spring Harbor, New York.

Robb, M. J., Wilson, M. A. & Vierula, P. J. (1995). A fungal actin-related protein involved in nuclear migration. *Molecular and General Genetics* **247**, 583–590.

Rosa, A. L., Alvarez, M. E. & Maldonado, C. (1990). Abnormal cytoplasmic bundles of filaments in the *Neurospora crassa snowflake* colonial mutant contain P59Nc. *Experimental Mycology* **14**, 372–380.

Scott, W. A. (1976). Biochemical genetics of morphogenesis in *Neurospora*. *Annual Review of Microbiology* **30**, 85–104.

Scott, W. A. & Abramsky, T. (1973). *Neurospora* 6-phosphogluconate dehydrogenase. II. Properties of two purified mutant enzymes. *Journal of Biological Chemistry* **248**, 3542–3545.

Scott, W. A. & Solomon, B. (1975). Adenosine 3',5'-cyclic monophosphate and morphology in *Neurospora crassa*: Drug induced alterations. *Journal of Bacteriology* **122**, 454–463.

Scott, W. A. & Tatum, E. L. (1970). Glucose-6-phosphate dehydrogenase and *Neurospora* morphology. *Proceedings of the National Academy of Sciences, U.S.A.* **66**, 515–522.

Springer, M. L. (1993). Genetic control of fungal differentiation: the three sporulation pathways of *Neurospora crassa*. *BioEssays* **15**, 365–373.

Steele, G. C. & Trinci, A. P. J. (1977). Effect of temperature and temperature shifts on growth and branching of a wild type and a temperature sensitive colonial mutant (*cot*-1) of *Neurospora crassa*. *Archives of Microbiology* **113**, 43–48.

Tatum, E. L., Barratt, R. W. & Cutter, V. M. Jr. (1949). Chemical induction of colonial paramorphs in *Neurospora* and *Syncephalastrum*. *Science* **109**, 509–511.

Temporini, E. D. & Rosa, A. L. (1993). Pleiotropic and differential phenotypic expression of two *sn* (snowflake) mutant alleles of *Neurospora crassa*: analysis in homokaryotic and heterokaryotic cells. *Current Genetics* **23**, 129–133.

Terenzi, H. F., Flawiá, M. M., Téllez-Iñón, M. T. & Torres, H. N. (1976). Control of *Neurospora crassa* morphology by cyclic adenosine 3',5'-monophosphate and dibutyryl cyclic adenosine 3',5'-monophosphate. *Journal of Bacteriology* **126**, 91–99.

Terenzi, H. F., Flawiá, M. M. & Torres, H. N. (1974). A *Neurospora crassa* mutant showing reduced adenylate cyclase activity. *Biochemical and Biophysical Research Communications* **58**, 990–996.

Wiebe, M. G., Robson, G. D. & Trinci, A. P. J. (1992). Evidence for the independent regulation of hyphal extension and branch initiation in *Fusarium graminearum* A3/5. *FEMS Microbiology Letters* **90**, 179–184.

Wrathall, C. R. & Tatum, E. L. (1974). Hyphal wall peptides and colonial morphology in *Neurospora crassa*. *Biochemical Genetics* **12**, 59–68.

Yarden, O., Plamann, M., Ebbole, D. J. & Yanofsky, C. (1992). *cot*-1, a gene required for hyphal elongation in *Neurospora crassa*, encodes a protein kinase. *EMBO Journal* **11**, 2159–2166.

Yarden, O. & Yanofsky, C. (1991). Chitin synthase I plays a major role in cell wall biogenesis in *Neurospora crassa*. *Genes and Development* **5**, 2420–2430.

Chapter 5

Nuclear changes during fungal development

SIU-WAI CHIU

Summary

This chapter examines whether nuclear changes during development can be generalized into a pattern for higher fungi. It covers the following topics: ploidy level, number of nuclei per cell, nuclear division and migration, and chromosome behaviour. Chromosome length polymorphisms are common in fungi. These changes in the life cycles of various fungi are reviewed. A sexual cell (basidium or ascus) can carry out either meiosis or mitosis or a modified chromosomal cycle. Meiosis does not require heterozygosity at the mating type locus. Most fungi are heterothallic but homothallic species/isolates do exist. The latter result from diploidy, presence of heterokaryotic nuclei in a meiotic spore or an unusual meiotic pathway.

Introduction

Unlike most animals and plants, fungi are able to carry out intranuclear mitoses but fungal nuclei are difficult to study because they are small and pleomorphic. Progress in understanding the nuclear changes during the fungal life-cycle has come from combining and integrating different approaches: biochemical, cytological and genetical. In addition to physiological properties, morphological characters and mating type factors as genetic markers, molecular markers have commonly been used to trace the movement of nuclei and segregation of chromosomes (Hintz, Anderson & Horgen, 1988; Cooley, 1992; Chiu, Kwan & Cheng, 1993; Kerrigan *et al.*, 1993; Rizzo & May, 1994). By comparison with the normal development of the wild type, molecular and cellular analysis of mutants has led to the characterization of specific mutational defects in sporulation processes (Kamada, Sumiyoshi & Takemaru, 1989; Clutterbuck, 1994), discovery of components involved and clarification of the role of these components in cytokinesis, nuclear division and nuclear movement as well (Doonan, 1992; Morris & Enos, 1992). Biochemical analysis reveals the metabolic pathways involved and their regulation. For instance, the mitosis or maturation promoting factor (MPF) has the properties of a protein kinase and is itself regulated by phosphorylation/dephosphorylation. These protein kinases and corresponding phosphatases directly govern the transitions of the cell cycle (Morris & Enos, 1992). Cytological studies reveal the fine microscopic details of nuclear behaviour (O'Donnell, 1994). Block and release experiments, on the other hand, establish the dependence relationships between sequential steps (Lu, 1982).

Spore germination, mitosis and cytokinesis

Spore germination

Spore germination often refers to the emergence of a germ tube with or without swelling of the spore. In the ascomycete *Aspergillus nidulans* which is regarded by many as a model filamentous fungus, a uninucleate G1-arrested conidium underwent its first nuclear division as the spore swelled to signal germination (Morris & Enos, 1992). The second nuclear division was generally concomitant with germ tube emergence. At the four nucleus stage, the fungus had already established a polarized growth axis. Then the most apical nucleus normally entered mitosis first and a wave of intranuclear mitotic activity passed down the hypha (Morris & Enos, 1992). In contrast, in the basidiomycete *Volvariella*, a germ tube emerged before the basidiospore nucleus migrated (Chiu, 1993). Therefore, the early events in spore germination cannot be generalized in fungi.

Coordination of mitotic activity and cytokinesis

Analysis of fungal mitosis has employed temperature-sensitive cell-division-cycle mutants and mutants defective in cytoskeleton structure, DNA-specific dyes to monitor number, distribution and conformation of nuclei within cells, and fluorochrome-conjugated antibodies raised against the spindle pole body (Butt *et al.*, 1989; Kamada *et al.*, 1989; Kamada, Hirai & Fujii, 1993; Morris & Enos, 1992; Clutterbuck, 1994; Harris, Morrell & Hamer, 1994). Spindle pole bodies are the centrosome-equivalent organelles which appear to regulate the polymerization of both the spindle and astral microtubules, and these, in turn, are the major effectors of mitosis.

Cytokinesis in filamentous fungi refers to formation of crosswalls, called septa, and occurs close to the site of nuclear division in many organisms (Valla, 1984). In dikaryotic mycelia of clamp-forming basidiomycetes there are two classes of postmitotic nuclear migration. Immediately after the completion of mitosis, the first three sibling nuclei migrate. This is followed by the migration of the fourth nucleus from the clamp cell into the subterminal hyphal segment. Different fungi may vary in the fine details of cytokinesis. In *Schizophyllum commune*, the septum separating the progeny nuclei in the parent hypha had formed before formation of the septum delimiting the clamp cell whereas in *Pleurotus ostreatus*, the formation of the clamp septum preceded that of the septum in the parent hypha (Yoon & Kim, 1994).

In addition, when cytokinesis and nuclear division are examined along the same time scale, the induced class I *sep* mutants in *Aspergillus nidulans* were defective in cytokinesis, failing to form septa and arresting at the third nuclear division stage. They also showed aberrant nuclear morphology (elongated and multilobed) (Harris *et al.*, 1994). Therefore, this class of mutants suggests the existence of a regulatory mechanism to ensure the continuation of nuclear division following the initiation of cytokinesis during spore germination. Nevertheless, karyokinesis and cytokinesis varied from a rather loose coupling in the hyphal stages where multinucleate cells were formed to strict coupling during spore formation where uninucleate conidial cells were produced (Trinci, 1978; Doonan, 1992).

Machinery for nuclear movement

Cytoskeleton structure has been observed to be associated with nuclear division, nuclear migration and cytokinesis (McKerracher & Heath, 1986; Osmani, Osmani & Morris, 1990; Clutterbuck, 1994). Treating a diploid fungus with an antimicrotubular drug, such as benomyl, induces haploidization (Anderson, Petsche & Franklin, 1984; Büttner *et al.*, 1994). Kamada and colleagues (1989 & 1993) induced and isolated benomyl-resistant (*ben*) mutants which were defective in structural genes for α- or β-tubulin in *Coprinus cinereus*. These *ben* mutants showed blockage of transhyphal migration of nuclei in dikaryosis (formation of dikaryotic hyphae with clamp cells) but had no effect on migration of nuclei into developing spores. This may indicate that the tubulin-mediated movement system may not be the sole mechanism for translocation of nuclei. Characterization of these *ben* mutants further revealed that microtubules participated in the pairing of the two conjugate nuclei (Kamada *et al.*, 1993).

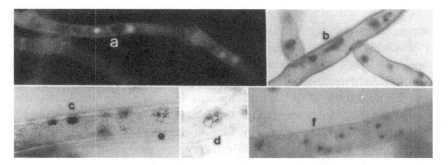

Fig. 1. Mitotic behaviour in fungi. a, dikaryotic nuclei; b, migrating interphase nucleus; c, prophase nucleus; d, metaphase nucleus; e, anaphase nucleus; f, telophase nuclei.

The involvement of actin, the other cytoskeleton component, has also been demonstrated (Grove & Sweigard, 1980; Doonan, 1992; Harris *et al.*, 1994). Immunohistochemical labelling of actin filaments reveals a close colocalization with the multiple nuclei in stem hyphae of *Flammulina* (Monzer, 1995). A contracting actin microfilament band, referred to as the 'septal band', appeared early at the position of a presumptive septum (Girbardt, 1979; Doonan, 1992; Kobori, Sato & Osumi, 1992). Inhibitor studies with cytochalasin A not only disturbed actin organization but also prevented septum formation (Grove & Sweigard, 1980; Kobori *et al.*, 1992). Gene disruption experiments further demonstrated the requirement for myosin, various phosphatases and actin-associated proteins for cytokinesis, cytoskeleton-cell wall attachment for nuclear migration as well as motility (Harris *et al.*, 1994; Kaminskyi & Heath, 1995).

Mitotic nuclear behaviour

In fungi, intranuclear mitotic nuclear behaviour with the nuclear membrane (= nuclear envelope) remaining intact or nearly so can be examined in living or preserved specimens after staining or using contrast-enhancing optics (Aist & Wilson, 1968; Taylor, 1985; Lü & McLaughlin, 1995). Fine details are revealed by electron microscopy of specimens prepared by conventional chemical fixation or freeze-substitution (Taylor, 1985; O'Donnell, 1994; Lü & McLaughlin, 1995). All cytological studies reveal similar events: resting interphase nuclei appear as spherical structures with diffuse chromatin while migrating interphase nuclei are characterized by the elongation of the nuclear envelope and presence of a nucleolus (Fig. 1).

With the duplication of the spindle pole body in S-phase (Taylor, 1985; Heath, 1994), the volume of the nucleus started to decrease and chromatin started to condense. The nucleus with a distinctive nucleolus, which was lost later at anaphase, enlarged until the progeny nuclei were formed. Frequent reports of the absence of a metaphase plate with highly condensed chromosomes may simply be a consequence of the very rapid progress through mitosis (Heath, 1994) (Fig. 1). Anaphase movement occurred at random directions to the long axis of the hypha (Aist & Wilson, 1968) (Fig. 1). At anaphase A, the chromatids moved to the poles of the spindle, whereas at anaphase B, the two poles moved further apart. At this stage, lagging chromosomes have often been reported (Kameswar, Aist & Crill, 1985; Lü & McLaughlin, 1995). Consequently, anaphase disjunction of chromatids is asynchronous. Heath (1994) summarized the 3 patterns observed at fungal telophase as follows: (i) median constriction so that the entire nucleoplasm is split in two and incorporated into the daughter nuclei; (ii) double constriction so that the daughter nuclei contain only a portion of the parental nucleoplasm, with a large portion being discarded into the cytoplasm and subsequently degraded via the cellular salvage pathways; and (iii) formation of a new envelope *de novo*, separate from the old envelope so that the chromosomes and a small portion of the nucleoplasm are incorporated into the daughter nuclei and the bulk of the parental nucleoplasm and envelope are discarded as in (ii).

In fungi with multinucleate vegetative thalli, mitotic nuclear division was usually asynchronous whereas in stem tissues, most nuclei were rapidly dividing and the degree of synchrony appeared to be increased (Doonan, 1992; Chiu, 1993).

Anastomosis and nuclear migration in vegetative mycelium

In mon-mon (monokaryon-monokaryon) matings, both unilateral and reciprocal nuclear exchange between the uninucleate parental thalli was common and was partially under the control of the mating type genes (Buller, 1931; Raper, 1966; May, 1988) whereas in pairings of homokaryotic *Agaricus* species which show a predominant haploid multinucleate life phase, compatible nuclei were only limited to the confrontation zone (Hintz *et al.*, 1988). Employing phase contrast photomicroscopic observations in the mon-mon pairing of *Schizophyllum commune*, it was found that the first formed heterokaryotic hyphae were in a transient multinucleate condition (Niederpruem, 1980a & b). Reduction in apical nuclear content leads to the stable dikaryotic condition at a later time and is by: (i) unequal segregation of nuclei following formation of a clamp connection (which will further become a clamp cell) and mitosis; (ii) long term entrapment of a single nucleus in the clamp connection as the predominant mechanism; and (iii) nuclear reduction via pseudoclamps.

In di-mon (dikaryon-monokaryon) mating of *Coprinus cinereus* and *Schizophyllum commune*, new dikaryotic associations were formed following nuclear migration into the monokaryon (Buller, 1931; Sweizynski & Day, 1960a & b; Raper, 1966). In diploid monokaryon pairings of *Armillaria* species which have a predominant diploid life phase, three types of nuclear inheritance were observed as revealed by molecular markers such as RFLPs (restriction fragment length polymorphism markers) and RAPD (random amplified polymorphic markers) (Rizzo & May, 1994; Carvalho, Smith & Anderson, 1995). They are: (i) the diploid nucleus replaced the haploid nuclei in most cases; or (ii) it maintained a stable 2N + N dikaryon without mitotic recombination or a triploid condition in rare cases; or (iii) mitotically recombinant nuclei were found.

However, for *in vitro* pairings between dikaryons and/or heterokaryons, rejection in terms of hyphal lysis usually occurred, and the phenomenon is named as vegetative or somatic or heterogenic incompatibility (Bégueret, Turrq & Clavè, 1994; Esser & Blaich, 1994). This is a complex situation which is under the control of multiple genes; nuclear replacement usually occurs, and this may serve as a means to preserve the genetic territory of the resident individual (Bégueret *et al.*, 1994; Esser & Blaich, 1994).

Meiosis

Meiotic nuclear behaviour

Detailed description of meiotic nuclear behaviour is available in the literature (Raju & Lu, 1973; Wells, 1977; Lu, 1993). Karyogamy took place rapidly and was immediately followed by zygotene. Meiotic nuclei were comparatively large (Figs 1 & 2). In contrast to plant and animal systems, fungal karyotype analysis is usually done at the pachytene stage at which chromosomes are paired but not at metaphase I at which chromosomes are at the maximum degree of condensation and appear as dots with conventional light microscopy. Tripartite synaptonemal (= synaptinemal) complexes were formed at pachytene of *Saccharomyces cerevisiae* and *Coprinus cinereus* but some fungi such as *Aspergillus nidulans*, *Podospora anserina* and *Schizosaccharomyces pombe* do not form synaptonemal complexes (Klein, 1994; Moens, 1994; Pukkila, 1994). At diplotene, the homologous

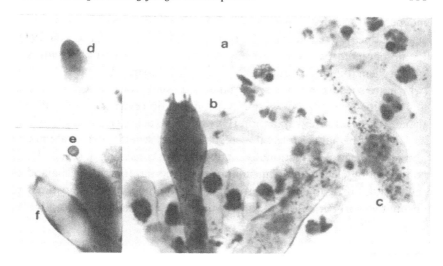

Fig. 2. Meiotic behaviour in fungi. a, prophase I nucleus; b, meiotic II division nuclei; c, interphase II nuclei; d, diakinesis nucleus; e, binucleate spore initial; f, empty basidium after spore discharge.

chromosome pairs appeared as lampbrush and fluffy. At meiotic I division, the nucleolus decreased in size while the spindle pole body divided. The meiotic II division was brief and resulted in tetrad nuclei each with a nucleolus (Fig. 2).

Meiosis takes place in the meiocytes (basidia or asci). This meiocyte developmental pathway has been dissected using multiple tools, including: (i) the high resolution *in situ* hybridization technique to define the time of spindle pole body separation as well as premeiotic DNA synthesis and to analyse the earliest stages of chromosome positioning as well as alignment, and (ii) the isolation of meiotic mutants to establish the landmarks of cell differentiation. These landmarks are: commitment to recombination, commitment to the first and second meiotic divisions, commitment to spore formation and commitment to spore maturation (Lu, 1982; McLaughlin, 1982; Ross & Margalith, 1987; Chiu & Moore, 1990; Padmore, Cao & Klechkner, 1991; Hawley & Arbel, 1993; Klein, 1994; Pukkila, 1994; see Chapter 6).

Until recently the central dogma of meiosis has been defined and stated as synapsis, recombination and segregation (Hawley & Arbel, 1993; Moens, 1994). It is now known that chromosome pairing and synapsis are mechanistically and temporally distinct processes (Padmore *et al.*, 1991). Premeiotic DNA replication and formation of synaptonemal complex are independent events (Kanda *et al.*, 1989 & 1990; Pukkila, 1994). Further, synapsis does not require homology (Hawley & Arbel, 1993; Pukkila, 1994). Homologous chromosomes are aligned prior to the earliest observation of synaptonemal complex and at this time double-strand break formation (which are sites for meiotic recombination) usually occurs. Using mutant analysis, it is found that DNA-DNA recombination events are not sufficient themselves to ensure disjunction (Hawley & Arbel, 1993; Pukkila, 1994). The synaptonemal complex, if formed, converts several sites of alignment and exchange into a functionally intact bivalent, consisting of kinetochores connected by chiasmata. Chiasmata serve the vital function of balancing the spindle forces on the kinetochores of homologous chromosomes to ensure their proper co-orientation at metaphase I (Hawley & Arbel, 1993; Moens, 1994).

Sexuality and meiosis

In higher fungi, heterothallism in which a haplont is self-sterile but cross-fertile with a compatible haplont is common (Ryvarden, 1991; Weber, 1992). Under suitable environmental stimulation formation of the spore distribution structure (the fruit body) takes place, and an abundant progeny is produced and dispersed for species propagation. In heterothallic fungi, karyogamy and meiosis pass the normal course typical for haplonts with the premeiotic replication of the two fusing nuclei occurring before karyogamy. In addition, many fungi show homokaryotic/monokaryotic fruiting (= haploid fruiting) (Raper, 1966; Arita, 1979; Weber 1992). In these cases, haploid meiosis or mitosis took place in basidia and, usually, few viable spores were produced (Raper, 1966; Zickler *et al.*, 1995). These haploid isolates possessing the capability of homokaryotic/monokaryotic fruiting usually resumed normal dikaryotic/heterokaryotic fruiting (the diploid fruiting) after mating. In *Podospora anserina*, a transgenic *mat* (mating type genes) mutant (strain originally deleted in the *mat* locus and later transformed with a mutagenized *mat* allele) bore asci carrying out meiosis in uninucleate crozier (Zickler *et al.*, 1995). Pachytene chromosomes often appeared as singles or paired irregularly but prophase nuclear evolution (nucleolar increased from leptotene to diplotene, and chromatin condensation occurred) and ascus growth were normal. The chromosomes were usually randomly distributed at anaphase I, resulting in aberrant numbers of chromosomes in the two second division nuclei. Abnormal spores, asci with uninucleate or one binucleate and two uninucleate spores were produced (Zickler *et al.*, 1995). The presence of this mutant indicates that karyogamy between two mutant nuclei of similar mating type also leads to meiosis and sporulation. Thus, heterozygosity of mating type is not a prerequisite for meiosis.

In fungi, homothallism is also encountered as exemplified by *Aspergillus nidulans* in which a haploid conidium can give rise to hyphae which are able to enter the sexual reproduction pathway. In *Urceolella tetraspora*, after karyogamy the fusion nucleus had only double the basic amount of DNA. Only 4 nuclei and 4 spores were formed (Weber, 1992). Unlike the normal meiocyte pathway, for some homothallic species, duplication of DNA content was not found at prekaryogamy stage but at prophase I (Oishi, Uno & Ishikawa, 1982; Bayman & Collins, 1990; Chiu, 1993; Chiu & Moore, 1993).

Petersen & Methven (1994) isolated some European clonal collections of heterothallic *Xerula radicata*. These strains did not show any mating reactions. Also, in their fruit bodies, hyphae were devoid of clamp connections, and only 2-spored basidia were produced (Petersen & Methven, 1994). Although the authors interpreted the European clone as monokaryotic fruiters, it may point to the evolution of homothallic strains from heterothallic fungi (Oishi *et al.*, 1982).

Postmeiotic mitosis and nuclear migration into meiotic spores

Some ascomycetes, such as *Ascobolus immersus*, *Neurospora crassa* and *Sordaria fimicola*, produce ordered tetrads (in which the position of the meiotic products reveals their origin as first or second division segregation) whereas the remaining ascomycetes and the basidiomycetes produce unordered tetrads. Many ascomycetes carry out one postmeiotic mitosis leading to the production of 8 ascospores per ascus. For basidiomycetes, there are many patterns. Fig. 3 summarizes the various patterns of nuclear behaviour in terms of meiosis, postmeiotic mitosis and nuclear migration, which has been observed in the homobasidiomycetes (Evans, 1959; Kühner, 1977; Chiu & Moore, 1993; Hibbett, Murakami & Tsuneda, 1994; Jacobson & Miller, 1994; Petersen, 1995).

In experiments performed *in vitro*, it was found that arrestment of meiosis and/or sporulation and modification of basidiospore number per basidium could be induced by mutation, chemical treatments, environmental conditions and the presence of lethal genes or spore killer elements (Lu, 1982; McLaughlin, 1982; Elliott & Challen, 1984; Kerrigan & Ross, 1987; Kanda *et al.*, 1989 & 1990; Chiu & Moore, 1990; Hasebe, Murakami & Tsuneda, 1991; Raju & Perkins, 1991). Assuming the random segregation of nuclei,

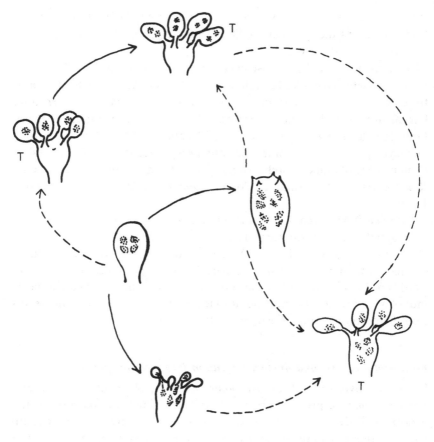

Fig. 3. Postmeiotic events in basidium. $--\rightarrow$, nuclear migration; $\rightarrow$, postmeiotic mitosis; T, terminal differentiated state.

Elliott & Challen (1983) used a mathematical approach to deduce a segregation ratio of 2:1 for heterokaryotic to homokaryotic progeny in secondary homothallic species which pack two compatible nuclei into the same spore. Yet in field collected isolates of secondarily homothallic *Agaricus bisporus* which actually bore 2-, 3- and 4-spored basidia (but at different frequencies), a significant deviation from the predicted ratio was observed (Kerrigan *et al.*, 1993). This observed deviation may be due to heterogeneous basidiospore numbers per basidium, the migration of one or two nuclei into basidiospores and the difference in germination rates of homokaryotic and heterokaryotic spores.

Many fungi show plasticity in life strategies as demonstrated in tetrapolar heterothallic *Suillus granulatus*; 5-10% basidiospores were binucleate spores which were secondarily homothallic as verified by RAPD markers (Jacobson & Miller, 1994). The germination rate and the presence of a functionally diverse mating system, resulting from heterothallic uninucleate spores (outbreeding) and secondarily homothallic binucleate spores, enable the flexibility in immediate reproduction and generation of genetic variation by outbreeding.

Ploidy level and lifecycles

Although the relative ploidy levels among different fungi has been deduced by comparing their DNA contents per nucleus, this method is found to be erratic. The genomic sizes of various haploid fungi vary widely, and within the same species, different strains may have different genome size (Tooley & Carras, 1992; Martin, 1995) (Table 1). Therefore, the estimation of ploidy level requires more sensitive methods such as southern hybridization of pulsed field gel resolving a fungal karyotype with a single copy chromosomal-specific probe (Barton & Gull, 1992; Büttner *et al.*, 1994), or comparison of DNA contents at various life cycle phases (Peabody, Motta & Therrien, 1978; Peabody & Peabody, 1986; Weber, 1992; Chiu, 1993; Chiu & Moore, 1993).

The basic DNA content of nuclei in the perfect (sexual; also called teleomorph) and imperfect (asexual; also called anamorph) stages of a fungus is in many cases the same. Reduction of DNA content per nucleus is usually limited to the transient diploid nucleus in meiocytes but in *Ascocryne sarcoides* and *Parorbiliopsis minuta*, the asexual conidial nuclei had only half the basic amount of DNA compared to the nuclei of apothecia (a special type of fruit body) and conidiophores (Weber, 1992). Therefore, a haploidization in the anamorphs was assumed (Weber, 1992). For multinucleate and clampless *Agaricus arvensis*, there was no significant difference in number of nuclei per cell of homo- and heterokaryons (Sonnenberg & Fritsche, 1989). In diploid *Armillaria* species, transient binucleate stages were found in several stages of the life-cycle: at initial stage of haploid mating, in some di-mon mating and in basidial initial except that of *A. mellea* (Tommerup & Broadbent, 1975; Peabody *et al.*, 1978; Watling, Kile & Gregory, 1982; Motta & Korhonen, 1986). When a fungus fruits, the number of nuclei per cell can be different from that at vegetative stage (Tables 2 & 3) (Tommerup & Broadbent, 1975; Wong & Gruen, 1977; Murakami & Takemaru, 1980; Gooday, 1985; Nakai, 1986; Chiu & Moore, 1993). In two *Armillaria* species, nuclei of the multinucleate tramal hyphae

Table 1. *Karyotypes of some fungi*

Fungus	Microscopy	PFGE	Genome (Mb)	Chromosome size (Mb)	Reference of PFGE
Agaricus bisporus	8, 9, 12	13	26.8; 34	1.2–3.5	Royer et al., 1992
Aspergillus nidulans	8	4 + 2 doublets	31	2.9–5.0	Brody & Carbon, 1989
Coprinus cinereus	4, 12, 13	13	ND	1.0–5.0	Pukkila & Skrzynia, 1993
Lentinula edodes	8	8	33	2.2–7.0	Arima & Morinaga, 1993
Neurospora crassa	7	3 + 2 doublets	47	4–12.6	Orbach et al., 1988
Phytophthora megasperma	13, 14; 17–23; 30–40	13; 9–10; ND	46.5; ND; ND	1.4–6.8; 2.5–6.1; 3.0–6.2	Tooley & Carras, 1992
Pleurotus ostreatus	8–10	6; 9; 10	20.8; 31.3; >39.5	2.1–5.2; 1.1–5.7; 2.3–>6	Sagawa & Nagata, 1992; Peberdy, Hanifah & Jia, 1993
Podospora anserina	7	5 + 1 doublet	34	3.8–6.0	Javerzat, Jacquier & Barreau, 1993
Saccharomyces cerevisiae	17	15 + 1 doublet	15	0.225–2.21	Schwartz & Cantor, 1984
Schizophyllum commune	8	4 + 2 doublets	35–36	1.2–3.5	Horton & Raper, 1991
Volvariella volvacea	9	5 + 4 doublets	34	1.3–5.4	Chiu et al., 1995

ND, not determined; PFGE, pulsed field gel electrophoresis.

Table 2. *Number of nuclei per stipe cell in some hymenomycetes*

Fungus	Range	Reference
Agaricus bisporus	4–32	Murakami & Takemaru, 1980
Armillaria mellea	1	Tommerup & Broadbent, 1975
Coprinus cinereus	2–156	Gooday, 1985
Flammulina velutipes	2–32	Wong & Gruen, 1977
Lentinula edodes	2–8	Nakai, 1986
Pholiota nameko	2–4	Murakami & Takemaru, 1980
Pleurotus pulmonarius	2–4	Chiu, unpublished
Volvariella bombycina	5–30	Chiu & Moore, 1993

and dikaryotic subhymenial cells were often haploid, indicating the presence of haploidization during fruiting before the normal haploidization process in meiocytes (Peabody *et al.*, 1978; Peabody & Peabody, 1985). The absence of a decreasing trend of the number of nuclei per cell indicates that there is no pre-selection mechanism of nuclei for karyogamy and meiosis in multinucleate fungi. In less frequent cases, nuclei with higher amounts of DNA occurred in fruit bodies (often in the tips of paraphyses, but also in all other types of cells) and endopolyploidy was therefore supposed (Evans, 1959; Weber, 1992). The ploidy level and number of nuclei per cell in various life phases of some fungi are shown in Table 3. Fungi seem to be very diverse and do not conform to a fixed pattern of change in ploidy level and nuclear number.

Unlike the normal dikaryotic fruit bodies, artificially synthesized triploid and tetraploid fruit bodies in *Coprinus cinereus* produced various abnormalities, including carrying out prolonged meiotic divisions or aborted meiosis, producing few viable spores or defective spores with 2 germ pores (Murakami & Takemaru, 1984). In triploid yeast *Saccharomyces cerevisiae*, triple-synapsis was common whereas synapsed bivalents instead of multiple synapsis were mostly encountered in tetraploid strains (Loidl, 1995). Similarly, in *Coprinus cinereus*, polyploid sexual basidia reduced spore viability and/or reduced normal spore production (Murakami & Takemaru, 1984; Loidl, 1995). Therefore, unlike higher plants, polyploidy is not common in higher fungi. Heterokaryon/dikaryon with a transient diploidy in meiocyte stage is the predominant pattern in the life-cycle of many fungi.

Table 3. Changes in number of nuclei in life-cycles of some hymenomycetes

Dominant life phase	sexual spore	Mono-/Homo-karyon	asexual spore	Di-/Hetero-karyon	stipe cell	lamellar trama	basidial initial	example	Reference
Haploid	1	1	nil	2	2	2	2	*Lentinula edodes, Pleurotus pulmonarus*	Chiu, unpublished
Haploid	2	1	1	2	multi-nucleate	2	2	*Flammulina velutipes*	Wong & Gruen, 1977
Haploid	2	1	1, 2	2	multi-nucleate	multi-nucleate	2	*Coprinus cinereus*	Gooday, 1985; Moore & Pukkila, 1985
Haploid	2	1-4	1, 2	2	2	2	1	*Pholiota nemako*	Arita, 1979
Haploid	2	multi-nucleate	ND	multi-nucleate	multi-nucleate	multi-nucleate	2	*Agaricus bisporus*	Evans, 1959
Haploid	1	1	ND	1	1	1	1	*Xerula radicata* European clones	Petersen & Methven, 1994
Diploid	1	1	ND	1	1	multi-nucleate	2	*Armillaria* complex	Tommerup & Broadbent, 1975
Dikaryon	2	ND	ND	2	ND	ND	2	*Coprinus bilanatus*	Ross & Margalith, 1987

Karyotype analysis

Lu (1993) developed a protocol which involves mechanical breakage of meiotic cells before spreading and staining with silver nitrate for the examination of synaptonemal complex and chromosomal re-arrangement by both light and electron microscopy. Scherthan *et al.* (1992) developed the fluorescence *in situ* hybridization (FISH) technique and applied it to chromosome spreads of *Saccharomyces cerevisiae*. With a probe derived from the homologous genomic library, the technique allows the study of specific individual chromosome during meiosis. Other techniques are pulsed field gel electrophoresis (PFGE) and southern hybridization. The karyotypes of most fungi can be resolved electrophoretically by PFGE (Table 1). Southern hybridization using homologous probes can reveal ploidy levels of different isolates (Barton & Gull, 1992; Büttner *et al.*, 1994) and gene amplification in development (Pukkila & Skrzynia, 1993). Further, this technique reveals the loss of supernumerary chromosomes and generation of novel-sized chromosomes by meiotic recombination even in homothallic fungi (Zolan, Heyler & Stassen, 1994; Carvalho *et al.*, 1995).

Conclusions

Fungal chromosomes are too minute and at the resolution limit of conventional light microscopy. Therefore, advanced instruments and techniques are required for observation and experimentation of fungal nuclear cytology. An integration of cytological, genetical and biochemical techniques in experimental studies continue to enable us understand the divergence and plasticity in nuclear changes during life-cycles of different fungi as their survival strategies.

Acknowledgements

SWC thanks UPGC for grants to perform some of the studies mentioned. She is grateful to Mr. Christopher Wing-Tat Chiu for getting some of the reference materials.

References

Aist, J. R. & Wilson, C. L. (1968). Interpretation of nuclear division figures in vegetative hyphae of fungi. *Phytopathology* **58**, 876-877.
Anderson, J. B., Petsche, D. M. & Franklin, A. L. (1984). Nuclear DNA content

of benomyl-induced segregants of diploid strains of the phytopathogenic fungus *Armillaria mellea*. *Canadian Journal of Genetics and Cytology* **27**, 47–50.

Arima, T. & Morinaga, T. (1993). Electrophoretic karyotype of *Lentinus edodes*. *Transactions of the Mycological Society, Japan* **34**, 481–485.

Arita, I. (1979). Cytological studies on *Pholiota*. *Reports of the Tottori Mycological Society (Japan)* **17**, 1–118. In Japanese with English summary.

Barton, R. C. & Gull, K. (1992). Isolation, characterization and genetic analysis of monosomic, aneuploid mutants of *Candida albicans*. *Molecular Microbiology* **6**, 171–177.

Bayman, P. & Collins, O. R. (1990). Meiosis and premeiotic DNA synthesis in a homothallic *Coprinus*. *Mycologia* **82**, 170–174.

Bégueret, J., Turrq, B. & Clavè, C. (1994) Vegetative incompatibility in filamentous fungi: *het* genes begin to talk. *Trends in Genetics* **12**, 441–446.

Brody, H. & Carbon, J. (1989). Electrophoretic karyotype of *Aspergillus nidulans*. *Proceedings of the National Academic Science, U.S.A.* **86**, 6260–6263.

Buller, A. H. R. (1931). *Researches on Fungi*. vol. 4. Longmans and Green: London.

Butt, T. M., Hoch, H. C., Staples, R. C. & St Leger, R. J. (1989). Use of fluorochromes in the study of fungal cytology and differentiation. *Experimental Mycology* **13**, 303–320.

Büttner, P., Koch, F., Voigt, K., Quidde, T., Risch, S., Blaich, R., Brckner, B. & Tudzynski, P. (1994). Variations in ploidy among isolates of *Botrytis cinerea*: implications for genetic and molecular analyses. *Current Genetics* **25**, 445–450.

Carvalho, D. B., Smith, M. L. & Anderson, J. B. (1995). Genetic exchange between diploid and haploid mycelia of *Armillaria gallica*. *Mycological Research* **99**, 641–647.

Chiu, S. W. (1993). Evidence for a haploid life-cycle in *Volvariella volvacea* by microspectrophotometric measurements and observation of nuclear behaviour. *Mycological Research* **97**, 1481–1485.

Chiu, S. W., Cheung, W. M. W., Chen, M. J., Chang, S. T. & Moore, D. (1995). Nucleotide sequence of 5.8S rDNA and electrophoretic karyotype of mushroom *Volvariella volvacea*. *Inoculum* **46**(3), 8.

Chiu, S. W., Kwan, H. S. & Cheng, S. C. (1993). Application of arbitrarily-primed polymerase chain reaction in molecular studies of mushroom species with emphasis on *Lentinula edodes*. In *Culture Collection and Breeding of Edible Fungi* (ed. S. T. Chang, J. A. Buswell & P. G. Miles), pp. 265–284. Gordon & Breach Pub. Inc.: Philadelphia.

Chiu, S. W. & Moore, D. (1990). Sporulation in *Coprinus cinereus*: use of *in vitro* assay to establish the major landmarks in differentiation. *Mycological Research* **94**, 249–253.

Chiu, S. W. & Moore, D. (1993). Cell form, function and lineage in the hymenia of *Coprinus cinereus* and *Volvariella bombycina*. *Mycological Research* **97**, 221–226.

Clutterbuck, A. J. (1994). Mutants of *Aspergillus nidulans* deficient in nuclear migration during hyphal growth and conidiation. *Microbiology* **140**, 1169–1174.

Cooley, R. N. (1992). The use of RFLP analysis, electrophoretic karyotyping and PCR in studies of plant pathogenic fungi. In *Molecular Biology of*

Filamentous Fungi (ed. U. Stahl & P. Tudzynski), pp. 13–25. VCH: Weinheim, New York, Basel & Cambridge.

Doonan, J. H. (1992). Cell division in *Aspergillus*. *Journal of Cell Science* **103**, 599–611.

Elliott, T. J. & Challen, M. P. (1983). Genetic ratio in secondary homothallic basidiomycetes. *Experimental Mycology* **7**, 170–174.

Elliott, T. J. & Challen, M. P. (1984). Effect of temperature on spore number in the cultivated mushroom, *Agaricus bisporus*. *Transactions of the British Mycological Society* **82**, 293–296.

Esser, K. & Blaich, R. (1994). Heterogenic incompatibility. In *The Mycota, I, Growth, Differentiation and Sexuality* (ed. J. G. H. Wessels & F. Meinhardt), pp. 211–232. Springer-Verlag: Berlin & Heidelberg.

Evans, H. J. (1959). Nuclear behaviour in the cultivated mushroom. *Chromosoma* **10**, 115–135.

Girbardt, M. (1979). A microfilamentous septal belt (FSB) during induction of cytokinesis in *Trametes versicolor* (L. ex Fr.). *Experimental Mycology* **3**, 215–228.

Gooday, G. W. (1985). Elongation of the stipe of *Coprinus cinereus*. In *Developmental Biology of Higher Fungi* (ed. D. Moore, L. A. Casselton, D. A. Wood, & J. C. Frankland), pp. 311–322. Cambridge University Press: Cambridge, U. K.

Grove, S. N. & Sweigard, J. A. (1980). Cytochalasin A inhibits spore germination and hyphal tip growth in *Gilbertella pensicaria*. *Experimental Mycology* **4**, 239–250.

Harris, S. D., Morrell, J. L. & Hamer, J. E. (1994). Identification and characterization of *Aspergillus nidulans* mutants defective in cytokinesis. *Genetics* **136**, 517–532.

Hasebe, K., Murakami, S. & Tsuneda, A. (1991). Cytology and genetics of a sporeless mutant of *Lentinus edodes*. *Mycologia* **83**, 354–359.

Hawley, R. S. & Arbel, T. (1993). Yeast genetics and the fall of the classical view of meiosis. *Cell* **72**, 301–303.

Heath, I. B. (1994). The cytoskeleton in hyphal growth, organelle movements and mitosis. In *The Mycota, I, Growth, Differentiation and Sexuality* (ed. J. G. H. Wessels & F. Meinhardt), pp. 43–65. Springer-Verlag: Berlin & Heidelberg.

Hibbett, D. S., Murakami, S. & Tsuneda, A. (1994). Postmeiotic nuclear behavior in *Lentinus, Panus* and *Neolentinus*. *Mycologia* **86**, 725–732.

Hintz, W. E. A., Anderson, J. B. & Horgen, P. A. (1988). Nuclear migration and mitochondrial inheritance in the mushroom *Agaricus bitorquis*. *Genetics* **119**, 35–41.

Horton, J. S. & Raper, C. A. (1991). Pulsed field gel electrophoretic analysis of *Schizophyllum commune* chromosomal DNA. *Current Genetics* **19**, 77–80.

Jacobson, K. M. & Miller, O. K. Jr. (1994). Postmeiotic mitosis in the basidia of *Suillus granulatus*: implications for population structure and dispersal biology. *Mycologia* **86**, 511–516.

Javerzat, J. P., Jacquier, C. & Barreau, C. (1993). Assignment of linkage groups to the electrophoretically-separated chromosomes of the fungus *Podospora anserina*. *Current Genetics* **24**, 219–222.

Kamada, T., Hirai, K. & Fujii, M. (1993). The role of the cytoskeleton in the pairing and positioning of the two nuclei in the apical cell of the dikaryon of the basidiomycete *Coprinus cinereus*. *Experimental Mycology* **17**,

338–344.

Kamada, T., Sumiyoshi, T. & Takemaru, T. (1989). Mutations in β-tubulin block transhyphal migration of nuclei in dikaryosis in the homobasidiomycete *Coprinus cinereus*. *Plant Cell Physiology* **30**, 1073–1080.

Kameswar, R. K. V. S. R., Aist, J. R. & Crill, J. P. (1985). Mitosis in the rice blast fungus and its possible implications for pathogenic variability. *Canadian Journal of Botany* **63**, 1129–1134.

Kaminskyj, S. G. W. & Heath, I. B. (1995). Integrin and spectrin homologues and cytoplasm-wall adhesion in tip growth. *Journal of Cell Science* **108**, 849–856.

Kanda, T., Arakawa, H., Yasuda, Y. & Takemaru, T. (1990). Basidiospore formation in a mutant of incompatibility factors and in mutants that arrest at meta-anaphase I in *Coprinus cinereus*. *Experimental Mycology* **14**, 218–226.

Kanda, T., Goto, A., Sawa, K., Arakawa, H., Yasuda, Y. & Takemaru, T. (1989). Isolation and characterization of recessive sporeless mutants in the basidiomycete *Coprinus cinereus*. *Molecular and General Genetics* **216**, 526–529.

Kerrigan, R. W. & Ross, I. K. (1987). Dynamic aspects of basidiospore number in *Agaricus*. *Mycologia* **79**, 204–215.

Kerrigan, R. W., Royer, J. C., Baller, L. M., Kohli, Y., Horgen, P. A. & Anderson, J. B. (1993). Meiotic behavior and linkage relationships in the secondarily homothallic fungus *Agaricus bisporus*. *Genetics* **133**, 225–236.

Klein, S. (1994). Choose your partner: chromosome pairing in yeast meiosis. *BioEssays* **16**, 869–871.

Kobori, H. Sato, M. & Osumi, M. (1992). Relationship of actin organization to growth in the two forms of the dimorphic yeast *Candida tropicalis*. *Protoplasma* **167**, 193–204.

Kühner, R. (1977). Variation of nuclear behaviour in the homobasidiomycetes. *Transactions of the British Mycological Society* **68**, 1–16.

Loidl, J. (1995). Meiotic chromosome pairing in triploid and tetraploid *Saccharomyces cerevisie*. *Genetics* **139**, 1511–1520.

Lu, B. C. (1982). Replication of deoxyribonucleic acid and crossing over in *Coprinus*. In *Basidium and Basidiocarp – Evolution, Cytology, Function and Development* (ed. K. Wells & E. K. Wells), pp. 93–112. Springer-Verlag: New York, Heidelberg and Berlin.

Lu, B. C. (1993). Spreading the synaptonemal complex of *Neurospora crassa*. *Chromosoma* **102**, 464–472.

Lü, H. & McLaughlin, D. J. (1995). A light and electron microscopic study of mitosis in the clamp connection of *Auricularia auricula–judae*. *Canadian Journal of Botany* **73**, 315–332.

Martin, F. N. (1995). Electrophoretic karyotype polymorphisms in the genus *Pythium*. *Mycologia* **87**, 333–353.

May, G. (1988). Somatic incompatibility and individualism in the coprophilous basidiomycete, *Coprinus cinereus*. *Transactions of the British Mycological Society* **91**, 443–451.

McKerracher, L. J. & Heath, I. B. (1986). Fungal nuclear behaviour analysed by ultraviolet microbeam irradiation. *Cell Motility & Cytoskeleton* **6**, 35–47.

McLaughlin, D. J. (1982). Ultrastructure and cytochemistry of basidial and basidiospore development. In *Basidium and Basidiocarp: Evolution, Cytology, Function and Development* (ed. K. Wells & E. K. Wells), pp. 37–73. Springer-Verlag: New York, Heidelberg and Berlin.

Moens, P. B. (1994). Molecular perspectives of chromosome pairing at meiosis. *BioEssays* **16**, 101–106.

Monzer, J. (1995). Actin filaments are involved in cellular graviperception of the basidiomycete *Flammulina velutipes*. *European Journal of Cell Biology* **66**, 151–156.

Moore, D. & Pukkila, P. J. (1985). *Coprinus cinereus*: an ideal organism for studies of genetics and developmental biology. *Journal of Biological Education* **19**, 31-40.

Morris, N. R. & Enos, A. P. (1992). Mitotic gold in a mold: *Aspergillus* genetics and the biology of mitosis. *Trends in Genetics* **8**, 32–37.

Motta, J. J. & Korhonen, K. (1986). A note on *Armillaria mellea* and *Armillaria bulbosa* from the middle Atlantic States. *Mycologia* **78**, 471–474.

Murakami, S. & Takemaru, T. (1980). Nuclear number in stipe cells of some hymenomycetes. *Reports of the Tottori Mycological Institute (Japan)* **18**, 143–148.

Murakami, S. & Takemaru, T. (1984). Cytological studies on basidiospore formation in polyploid fruitbodies of *Coprinus cinereus*. *Report of the Tottori Mycological Institute (Japan)* **22**, 67–68.

Nakai, Y. (1986). Cytological studies on shiitake, *Lentinus edodes* (Berk.) Sing. *Reports of the Tottori Mycological Institute (Japan)* **24**, 1–202.

Niederpruem, D. J. (1980a). Direct studies of dikaryotization in *Schizophyllum commune*. I. Live intercellular nuclear migration patterns. *Archives of Microbiology* **128**, 172–178.

Niederpruem, D. J. (1980b). Direct studies of dikaryotization in *Schizophyllum commune*. II Behavior and fate of multikaryotic hyphae. *Archives of Microbiology* **128**, 172–178.

O'Donnell, K. (1994). A reevaluation of the mitotic spindle pole body cycle in *Tilletia caries* based on freeze-substitution techniques. *Canadian Journal of Botany* **72**, 1412–1423.

Oishi, K., Uno, I. & Ishikawa, T. (1982). Timing of DNA replication during the meiotic process in monokaryotic basidiocarps of *Coprinus macrorhizus*. *Archives of Microbiology* **132**, 372–374.

Orbach, M. J., Vollrath, D., Davis, R. W. & Yanofsky, C. (1988). An electrophoretic karyotype of *Neurospora crassa*. *Molecular and Cellular Biology* **8**, 1469–1473.

Osmani, A. H., Osmani, S. A. & Morris, N. R. (1990). The molecular cloning and identification of a gene product specifically required for nuclear movement in *Aspergillus nidulans*. *Journal of Cell Biology* **111**, 543–552.

Padmore, R., Cao, L. & Klechkner, N. (1991). Temporal comparison of recombination and synaptonemal complex formation during meiosis in *Saccharomyces cerevisiae*. *Cell* **66**, 1239–1256.

Peabody, D. C., Motta, J. J. & Therrien, C. D. (1978) Cytophotometric evidence for heteroploidy in the life cycle of *Armillaria mellea*. *Mycologia* **70**, 487–498.

Peabody, D. C. & Peabody, R. B. (1986). Nuclear volumes and DNA content of three stages in the life cycle of *Armillaria bulbosa*. *Mycologia* **78**, 967–968.

Peberdy, J. F., Hanifah, A. H. & Jia, J-H. (1993). New perspectives on the genetics of *Pleurotus*. In *Mushrom Biology and Mushroom Products* (ed.

S-T. Chang, J. A. Buswell & S. W. Chiu), pp. 55–62. The Chinese
University Press: Hong Kong.

Petersen, R. H. (1995). There's more to a mushroom than meets the eye: mating
studies in the agaricales. *Mycologia* **87**, 1–17.

Petersen, R. H. & Methven, A. S. (1994). Mating systems in the Xerulaceae:
Xerula. Canadian Journal of Botany **72**, 1151–1163.

Pukkila, P. J. (1994). Meiosis in mycelial fungi. In *The Mycota, I, Growth,
Differentiation and Sexuality* (ed. J. G. H. Wessels & F. Meinhardt), pp.
267–281. Springer-Verlag: Berlin, Heidelberg.

Pukkila, P. J. & Skrzynia, C. (1993). Frequent changes in the number of
reiterated ribosomal RNA genes throughout the life cycle of the
basidiomycete *Coprinus cinereus. Genetics* **133**, 203–211.

Raju, N. B. & Lu, B. C. (1973). Meiosis in *Coprinus.* IV Morphology and
behaviour of spindle pole bodies. *Journal of Cell Biology* **12**, 131–141.

Raju, N. B. & Perkins, D. D. (1991). Expression of meiotic drive elements spore
killer-2 and spore killer-3 in asci of *Neurospora tetrasperma. Genetics* **129**,
25–37.

Raper, J. R. (1966). *Genetics of Sexuality of Higher Fungi.* The Ronald Press
Co.: New York.

Rizzo, M. & May, G. (1994). Nuclear replacement during mating in *Armillaria
ostoyae* (Basidiomycotina). *Microbiology* **140**, 2115–2124.

Ross, I. K. & Margalith, P. (1987). Nuclear behaviour in the basidia of the
secondarily homothallic *Coprinus bilanatus. Mycologia* **79**, 595–602.

Royer, J. C., Hintz, W. E., Kerrigan, R. W. & Horgen, P. A. (1992).
Electrophoretic karyotype analysis of the button mushroom, *Agaricus
bisporus. Genome* **35**, 694–698.

Ryvarden, L. (1991). *Genera of Polypores: Nomenclature and Taxonomy.
Synopsis Fungorum* **5**, 1–363.

Sagawa, I. & Nagata, Y. (1992). Analysis of chromosomal DNA of mushrooms
in genus *Pleurotus* by pulsed field gel electrophoresis. *Journal of General
and Applied Microbiology* **38**, 47–52.

Scherthan, H., Loidl, J., Schuster, T. & Schweizer, D. (1992). Meiotic
chromosome condensation and pairing in *Saccharomyces cerevisiae* studied
by chromosome painting. *Chromosoma* **101**, 590–595.

Schwartz, D. C. & Cantor, C. R. (1984). Separation of yeast chromosome-sized
DNAs by pulsed field gel electrophoresis. *Cell* **37**, 67–75.

Sonnenberg, A. S. & Fritsche, G. (1989). Cytological observations in *Agaricus
arvensis. Mushroom Science* **12**(I), 101–107.

Swiezynski, K. M. & Day, P. R. (1960a). Heterokaryon formation in *Coprinus
lagopus. Genetical Research* **1**, 114–128.

Swiezynski, K. M. & Day, P. R. (1960b). Migration of nuclei in *Coprinus
lagopus. Genetical Research* **1**, 129–139.

Taylor, J. W. (1985). Mitosis in the basidiomycete fungus *Tulasnella araneosa.
Protoplasma* **126**, 1–18.

Tommerup, I. C. & Broadbent, D. (1975). Nuclear fusion, meiosis and the
origin of dikaryotic hyphae in *Armillaria mellea. Archives of Microbiology*
103, 279–282.

Tooley, P. W. & Carras, M. M. (1992). Separation of chromosomes of
Phytophthora megasperma species using gel electrophoresis. *Experimental
Mycology* **16**, 188–196.

Trinci, A. P. J. (1978). The duplication cycle and vegetative development in moulds. In *The Filamentous Fungi*, **Vol. 3**, *Developmental Mycology* (eds. Smith, J. E. & Berry, D. R.), pp. 132–163. Edward Arnold: London.

Valla, G. (1984). Changes in DNA content of nuclei in apical and intercalary compartments of *Polyporus arcularius* during hyphal growth. *Transactions of the British Mycological Society* **83**, 265–273.

Watling, R., Kile, G. A. & Gregory, N. M. (1982). The genus *Armillaria* - nomenclature, typification, the identity of *Armillaria mellea* and species differentiation. *Transactions of the British Mycological Society* **78**, 271–285.

Weber, E. (1992). Untersuchungen zu Fortpflanzung und Ploidie verschiedener Ascomyceten. *Bibliotheca Mycologica* **140**, 1–186. In German with English summary.

Wells, K. (1977). Meiotic and mitotic divisions in the basidiomycotina. In *Mechanisms and Control of Cell Division* (ed. T. L. Rost & E. M. Gifford, Jr.), pp. 337–374. Dowden, Hutchinson & Ross: Strougsburg, PA, U.S.A.

Wong, W. M. & Gruen, H. E. (1977). Changes in cell size and nuclear number during elongation of *Flammulina velutipes* fruitbodies. *Mycologia* **69**, 899–913.

Yoon, K. S. & Kim, Y. S. (1994). Ultrastructure of mitosis and clamp formation in the somatic hyphae of *Pleurotus ostreatus*. *Mycologia* **86**, 593–601.

Zickler, D., Arnaise S., Coppin, E., Debuchy, R. & Picard, M. (1995). Altered mating-type identity in the fungus *Podospora anserina* leads to selfish nuclei, uniparental progeny and haploid meiosis. *Genetics* **140**, 493–503.

Zolan, M. E., Heyler, N. K. & Stassen N. Y. (1994). Inheritance of chromosome-length polymorphisms in *Coprinus cinereus*. *Genetics* **137**, 87–94.

Chapter 6

Experimental approaches to the study of pattern formation in *Coprinus cinereus*

ADRIAN N. BOURNE, SIU-WAI CHIU AND DAVID MOORE

Summary

The basidiomycete *Coprinus cinereus* has some special features which particularly suit it to experimental studies. In addition to ease of culture, which makes conventional approaches with various cytological techniques (including continuous video observation) practicable, we show in this chapter how a number of transplantation and bioassay experiments have been used to identify the sequence of commitment steps culminating in the final differentiated stage of sporulation in the meiocyte (basidium). Multicellular fruit bodies have the functions to produce, protect, support and help dispersal of the spores. We present here various case studies which have demonstrated the morphogens in control of the fruit body phenotype and identified the independent but co-ordinated morphogenetic subroutines which form a normal fruit body. These experimental approaches to agaric developmental biology enable formulation of various working models of fungal cell differentiation, tissue patterning and fruit body morphogenesis, thus providing a unifying theme for categorising fruit body ontogeny and for clarifying phylogenetic and taxonomic relationships.

Introduction

Although study of fungal developmental biology is not as flourishing as similar studies with plant and animal systems, a number of approaches have been pioneered in *Coprinus* which could be more widely applied. In addition to continuous monitoring and measurement on fruiting development (see Chapter 1), these approaches include surgical intervention in fruit body morphogenesis, explantation of fruit body tissues to defined media *in vitro*, use of metabolic inhibitors to interfere with development, study of mutants with defects in known developmental processes and use of tropic responses as models of morphogenesis (see Chapter 7).

126

Coprinus cinereus fruits readily on horse dung manure or semi-synthetic media within 14 to 21 days under controlled conditions (Anderson, 1971; Morimoto & Oda, 1973; Morimoto, Suda & Sagara, 1981; Moore & Pukkila, 1985). This fungus shows a typical haploid monokaryon – dikaryotic life cycle (Chiu & Moore, 1993; see Chapter 5). The genus *Coprinus* is unique because it carries out synchronous meiosis. This makes it a very convenient subject for study of the meiotic division both cytologically and biochemically (Lu, 1982; Montgomery & Lu, 1990; Moore & Pukkila, 1985; Pukkila, Shannon & Skrzynia, 1995; Zolan *et al.*, 1995). Progress in the meiocyte pathway can be manipulated by environmental factors such as light and temperature (Lu & Chiu, 1978; Moore, Horner & Liu, 1978; Kamada, Kurita & Takemaru, 1978) or addition of chemicals such as hydroxyurea (Raudaskoski & Lu, 1980; Lu, 1982; Chiu & Moore, 1988b). This is also a good experimental tool for conventional genetic and cytogenetic as well as molecular analyses (Moore & Pukkila, 1985; Casselton *et al.*, 1995; see Chapter 5) since it produces abundant oidia in monokaryons as well as sexual homokaryotic binucleate basidiospores. Both types of spores germinate readily on conventional agar media.

Surgery and explantation

Mycologists expect to be able to recover vegetative mycelial cultures from the tissues of fruit bodies (and other multicellular structures) collected in the field. This expectation is more often fulfilled than not, and usually with quite simple media. In addition, it is frequently possible to find appropriate environmental and nutritional conditions in which to cultivate the result-ant isolate to reform the fruit body *in vitro*. Neither plant nor animal scientists can contemplate such routine preparation of cell cultures from excised slivers of fully differentiated tissues, still less the regeneration of the whole organism. Such experience raises the question whether fungal multicellular structures consist of cells as fully committed to a differentiated state as are their plant or animal counterparts (see Section entitled *Fuzzy logic in fungal differentiation* in Chapter 1), but here we wish to emphasise the processes which do continue in explanted tissues rather than those which default to the vegetative state.

Renewed fruiting

Renewed fruiting is a phenomenon distinct from regeneration. In it, fragments of a fruit body explanted to a nutrient-rich or even water agar medium produce a new crop of fruit bodies with unusual rapidity instead of regenerating the missing tissues. Such rapid formation of fruit body primordia on excised fruit body tissues is not uncommon in agarics, but the high degree of developmental synchrony characteristic of the smaller *Coprinus* species permits assessment of its dependence on the physiological state of the tissue. The phenomenon was systematically investigated in *Coprinus congregatus* by Bastouill-Descollonges & Manachère (1984) who used the phrases '. . . this potential for direct regeneration . . . remains "memorized" in the inocula' and '. . . the competence of hymenial lamellae to sporulate in an autonomous way . . .', clearly implying their belief that fruit body tissues used as inocula in such experiments are in some way committed or channelled towards fruit body construction. Whether this represents some sort of cytological memory (and see Chapter 2) has not been clearly established. However, we have verified that explanted fruit body fragments of *C. cinereus* behave similarly (Chiu & Moore, 1988a).

Renewed fruiting is described as 'direct' when the new primordia are formed only on the original inoculum; as 'indirect' when primordia form only on the outgrowing mycelium; or as 'mixed' when primordia form on both inoculum and mycelium. All cultures inoculated with the basal portions of stems from normally-grown fruit bodies (16 samples) and all parts of the pseudorhizas (= extended stem bases) of dark-grown fruit bodies (10 samples) made direct fruit bodies within 4 d. This compares with cultures inoculated with vegetative dikaryon which, under the same conditions, formed fruit bodies in 10–14 d. The other portions of stems of normally-grown fruit bodies (16 samples) produced (in 9–12 d) all types of fruiting pattern, unpredictably from their physiological age or physical size at the time of inoculation. The types of fruiting pattern observed on cultures inoculated with isolated gills are summarised in Table 1. This shows a clear dependence of the type of fruiting pattern on physiological age; tissues explanted prior to karyogamy showing a preponderance of direct fruiting, those explanted during or after meiosis showing a minimum of direct fruiting. Exactly similar results were obtained for *C. congregatus* by Bastouill-Descollonges & Manachère (1984).

Fruit body tissues used as inocula are clearly much more competent to initiate fruiting than the average vegetative dikaryon inoculum but the circumstances which confer this competence are completely obscure. The only physiological aspect of normal fruit body development with which a correlation might be evident is the disposition of accumulated glycogen. Intracellular glycogen features prominently in various aspects of growth and development in *Coprinus*, including the vegetative mycelium (Madelin, 1960; Jirjis & Moore, 1976), sclerotium maturation (Waters, Moore &

Table 1. *Fruiting pattern of cultures inoculated with gills of* Coprinus cinereus

Physiological age at explantation	Number of cultures	Fruiting pattern observed (% of cultures)		
		Direct*	Indirect	Mixed
Dikaryotic (prekaryogamy)	19	63	32	5
Prophase I	38	24	37	39
Meiotic division	25	8	32	60
Sporulation	39	18	26	56

* see text for definitions. Data from Chiu & Moore, 1988a.

Butler, 1975b) and fruit body development (Blayney & Marchant, 1977; Moore, Elhiti & Butler, 1979; Gooday, 1985; Moore, Liu & Kuhad, 1987). The developing fruit body of *C. cinereus* accumulates large quantities of glycogen, which appear first in the stem base and later in the subhymenial regions of the gills (Moore *et al.*, 1979; Gooday, 1985). The greatly reduced frequency of direct fruiting in cultures initiated with lamellae explanted during or after meiosis (Table 1) seems to correlate with the rapid, immediately post-meiotic, utilization of glycogen (Moore *et al.*, 1987). Whether intracellular glycogen serves as a nutrient or whether this carbohydrate accumulation represents a general nutritional/developmental commitment for fruiting is worth testing. Brunt & Moore (1989) claimed that the yield of fruit bodies arising directly on the initial inoculum showed a positive correlation with the glycogen content of the culture inoculum. The relationship between direct fruiting and glycogen content of the inoculum was complex and could only be fitted by third-degree polynomial regression. Meanwhile, supplementation of the medium with commercial rabbit liver glycogen had no effect. In *Pleurotus sajor-caju* (= *P. pulmonarius*), no correlation could be found between glycogen content and fruiting capacity in experiments featuring *in vivo* fruiting on media with different carbon sources and *in vitro* renewed fruiting of excised stems (Chiu & To, 1993). Subsequent detailed analysis of the relation between glycogen concentration and fruit body development in *C. cinereus* concluded that the carbohydrate cannot be linked exclusively, or even predominantly, with any one of the several processes during fruit body maturation pathway (Ji & Moore, 1993). Therefore, the endogenous glycogen level does not represent a fruiting signal.

In *C. cinereus*, cAMP has been shown to activate glycogen phosphorylase and inhibit glycogen synthetase via the cAMP-dependent protein kinase *in vitro* (Uno & Ishikawa, 1981, 1982). Yet the addition of femtomolar concentrations of cAMP to vegetative cultures stimulated glycogen synthesis (Kuhad, Rosin & Moore, 1987). Fruiting ability was also correlated with high cAMP level in the data of Swamy, Uno & Ishikawa (1985). However, in these experiments fruit body tissues were not separated from the mycelium prior to analysis, and when this was done in *C. cinereus* it was shown that though cAMP concentration was elevated in the fruit body initial, it dropped to a level approaching that of the vegetative mycelium during meiosis and sporulation (Kuhad *et al.*, 1987). Therefore, cAMP is an unlikely candidate as a fruiting signal molecule.

Fruit body surgery

Most surgical experiments on fruit bodies *in vivo* have been carried out on *Flammulina*, *Coprinus* and *Agaricus* with a view to studying the possible involvement of hormones in fruit body development (see Chapter 7). Here, we wish only to mention some experiments which demonstrate that nutrient translocation (as opposed to hormone flow) in *C. cinereus* occurred mainly, but not exclusively, in the 'stem to cap' direction (Ji & Moore, 1993; Chapter 1).

Surgical treatments were performed on both caps and stems to try to block potential translocation routes. Small pieces of aluminium foil were inserted into tangential incisions in caps of fruit bodies 4.5 h after karyogamy. The fruit bodies were then kept at 26° in an illuminated incubator and glycogen content in the cap was measured every 2 h. After 2 h the glycogen concentrations in the surgically treated and untreated segments were not significantly different; after 4 h the surgically treated segment contained significantly less glycogen than the untreated segment; and 6 h after surgery there was, again, no significant difference between the two segments. When the same type of surgery was performed on fruit bodies in which basidiospores were already pigmented (about 11 hours after karyogamy) there was no significant difference between the segments after 1.5 h but after 3 and 4.5 h further incubation the part of the cap beneath the inserted aluminium foil had a significantly lower glycogen content than the untreated part of the cap. In the young fruit bodies in which the margins of primary gills are connected to the stem (tramal hyphae interweaving with the stem hyphae), a transient reduction in glycogen content seemed to be compensated so that 6 h after surgery the glycogen content of the cap below the incision was not significantly different from the control sample. No such compensation occurred in older fruit bodies whose tramal connections were broken. The results of these surgical experiments consequently imply that materials could be translocated between the gills or between stem and gill in the young cap, so bypassing the incision and aluminium foil insert. Detached from the stem, the older cap was limited to radial translocation routes (i.e. from apex to margin) through the cap flesh and these were successfully blocked by the insert. In other words, net translocation of glycogen from stem to cap is indicated, continuing well beyond formation and pigmentation of basidiospores.

Thus, there must be an organised intracellular nutrient translocation circuit from mycelium to stem and to cap. However, there is reason to believe that the stem receives material from the cap, too. Hammad *et al.* (1993) demonstrated that stem elongation benefits considerably from the presence of the cap. Intact fruit bodies elongated about 25% more than decapitated ones, this amounting to 2 to 3 cm greater length (see Table 2 in Chapter 1). Furthermore, while the major supply route to the gills runs radially from the cap apex through the cap flesh, in young fruit bodies gills can be serviced by translocation of nutrients through the connections which exist between the stem and the 'edges' of primary gills. Taken together, these observations indicate that a fruit body is provided with sophisticated translocation flows in all directions throughout its structure. Translocation of growth factors through these translocation routes may help in defining the morphogenetic pattern of the developing structure (see Chapter 7).

Explantation experiments

The ultimate surgical intervention is complete removal of a segment of tissue from its place *in vivo* for transfer to a cultivation medium *in vitro* - this we describe as an explantation experiment. In classical (animal) embryology such a procedure is a test for the level of commitment of the explanted tissue (see discussion in Slack, 1983). If the transplanted cell continues along the developmental pathway characteristic of its origin then it is said to have been committed prior to transplant. On the other hand, if the transplanted cell embarks upon the pathway appropriate to its new environment then it was clearly not committed at the time of transplant. Most fungal structures produce vegetative hyphae very readily when disturbed and 'transplanted' to a new 'environment' or medium. Most, but not all.

Commitment in the *Coprinus* hymenium was demonstrated in *C. cinereus* by McLaughlin (1982), and in *C. congregatus* by Bastouill-Descollonges & Manachère (1984). However, these authors did not discuss their experiments from this viewpoint, placing more stress in the former case on

Fig. 1. Scanning electron micrographs showing the hymenial cell types of *Coprinus cinereus*. a, paraphysis; b, long basidium; c, short basidium; d, basidiospore; and e, the cystidium-cystesium pair spanning between two neighbouring gills. Bar = 20 μm.

sterigma formation, and in the latter on the potential for renewed fruiting from excised lamellae (see above). Detailed analysis of commitment in *C. cinereus* using tissue explantation was done by Chiu & Moore (1988a). In this study commitment to hymenium development was determined by cytological examination of specimens of explanted gills.

In gills explanted at the dikaryotic stage (prior to meiosis) the majority of young basidia were arrested in differentiation at the stage which they had reached at the time of removal from the parent fruit body, even though the 2 day incubation period was sufficient to permit hyphal outgrowths to be formed, largely from tramal tissues of the gill. Young basidia of gill samples taken at or after prophase I all completed meiosis and sporulation after explantation (25 specimens). In contrast, paraphyses and cystidia (Fig. 1) in the same samples reverted to hyphal growth by unipolar or multipolar hyphal apex formation and/or continued to swell into giant cells. Thus, determination to sporulation was demonstrable, but only in materials explanted at prophase I or later. This is similar to the situation in *Saccharomyces cerevisiae*, where commitment to recombination does not inevitably lead to commitment to meiotic division, the latter requiring duplication of the spindle pole body which occurs early in the first meiotic division (Berry, 1983; Dawes, 1983). Raju & Lu (1973) found that the spindle pole body duplicated at diplotene in *C. cinereus*. Thus, at least on this basis, *S. cerevisiae* and *C. cinereus* seem to share similar requirements for the attainment of competence and commitment to recombination and meiotic division.

A difference was evident, however, in that *S. cerevisiae* cells removed from sporulation medium after commitment to recombination but before commitment to sporulation were able to return to mitotic vegetative growth (Berry, 1983). In the experiments with *C. cinereus*, all the isolated gills which were explanted at the dikaryotic stage maintained their hymenial structure even after 2 days incubation; hyphal outgrowths which did occur penetrated through the hymenium from the tramal tissues below. Thus, although such young basidia were unable to continue development on explantation; they were somehow inhibited from reversion to the vegetative state; i.e. they were specified as meiocytes but not yet determined for sporulation.

All the evidence suggests that prophase I is the critical stage at which *C. cinereus* basidia become determined for the division programme. Similar results were obtained whether water agar, buffered agar or nutrient agar was used as explantation medium. Meiosis in basidia, once initiated, was endogenously regulated and proceeded autonomously. The autonomous, endotrophic phenomenon and the synchrony of nuclear division in *C. cinereus* make isolated gills and stems ideal subjects for *in vitro* bioassay to study stem morphogenesis and basidial differentiation (Chiu & Moore, 1988b, 1990b; Moore, Hammad & Ji, 1994; see Chapters 1 and 7).

Using metabolic inhibitors

Dissection of the basidial pathway

Coprinus cinereus is an ammonia fungus, favouring ammonia-rich substrata, and naturally fruits on animal manure or straw enriched with urea. Therefore, the first experiment using the explantation bioassay examined the effect of ammonium and found that ammonium ions halted meiocyte differentiation; sporulation being terminated, with vegetative hyphae emerging from those parts of the basidium which were in active growth at the time of exposure (Chiu & Moore, 1988b). Tests at various pH values and ammonium (chloride and sulphate) concentrations showed that highly alkaline pH values inhibited gill development, but at permissive pH values (6-8) ammonium concentrations of 50 mM were inhibitory. Neither potassium chloride nor potassium sulphate had any effect. Two ammonium analogues, hydroxylamine and methylamine, were also effective in inhibiting sporulation *in vitro*. Other compounds, such as L-glutamine (but not D-glutamine) and L-methionine, were effective inhibitors, too (Chiu & Moore, 1988b). In contrast, two utilizable sugars (D-glucose, D-fructose), mannitol, and most metabolites of the tricarboxylic cycle, arginine cycle and urea cycle which are amplified during cap maturation (Ewaze, Moore & Stewart, 1978) were ineffective (Chiu & Moore, 1988b). Similarly, in *Saccharomyces cerevisiae*, L-methionine, ammonium and L-glutamine were also inhibitors of sporulation (Miller, 1963; Piñon, 1977; Delavier-Klutchko *et al.*, 1980; Freese *et al.*, 1984).

Ammonium salts injected into the caps of young fruit bodies with a microsyringe also terminated further meiocyte development. Very young primordia (prekaryogamy) were not able to withstand the damage caused by injection and in most cases aborted. Injections of 2.5 μl of 1 M ammonium salt solutions (buffered to pH 7) were effective in locally suppressing sporulation if injected in post-meiotic and early sporulation stages. White zones appeared around the point of injection as the rest of the cap matured and produced its crop of blackened spores. Similar injections of water or buffer had no visible effect on fruit body maturation. Ammonium ions inhibited the meiocyte development pathway *in vitro* when applied at any time during meiosis (stages prophase I through to the second meiotic division were tested). When applied at similar stages *in vivo*, ammonium retarded the rate of progress through meiosis but did not suppress sporulation. When applied at later sporulation stages (sterigma formation, spore formation, spore pigmentation), ammonium arrested sporulation completely both *in vivo* and *in vitro*.

Cytological examination of gills excised at prophase I and explanted to ammonium-supplemented medium for 24 h showed a range of responses. Some were arrested at prophase I, others continued to metaphase I and some even completed the meiotic division, but no sporulation was observed. Although meiosis is well synchronised in *Coprinus*, synchrony is not perfect and the different stages at which development was arrested presumably reflected a combination of variation in exact time of exposure to ammonium and variation in stage reached by the time of exposure. Samples which were explanted at later stages suffered ammonium-arrest at correspondingly later stages. Tissue taken during meiosis (prophase I, meiotic divisions I and II) showed basidia arrested in later meiotic stages and in early sporulation stages. However, tissue explanted during those early sporulation stages seemed to become arrested immediately. Thus, exposure to ammonium caused termination of the normal developmental sequence of the basidium (Fig. 2).

Chiu & Moore (1988b) used a standard time of exposure in their bioassay. Moore *et al.* (1994) modified the technique to establish the time of exposure to 80 mM NH_4Cl required for inhibition of sporulation, and the

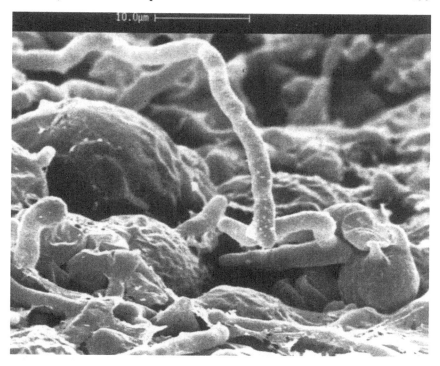

Fig. 2. Scanning electron micrograph showing a typical stress response from a basidium in comparison to a giant sterile hymenial cell and hyphae penetrating from the gill trama into the hymenium of an excised gill.

most sensitive time period during the course of meiosis. The time of exposure required varied directly with the stage in meiosis in the tissue at the time of excision. Tissue excised as karyogamy was occurring required 7 to 8 h exposure to NH_4Cl for sporulation to be halted, tissue excised during meiotic division I required only 2 to 3 h exposure. Observations made with 16 different fruit bodies established that the stage just after completion of the second meiotic division but before the appearance of sterigmata (spanning 60 to 90 min) was most sensitive to inhibition by ammonium. In all of these bioassays (Chiu & Moore, 1988a; Moore *et al.*, 1994), ammonium causes the rapid and regular promotion of reversion to hyphal tip growth from basidia, breaking down the commitment normally shown to their developmental pathway.

The demonstration that sterile elements of the hymenium immediately revert to hyphal growth on explantation to agar media (Chiu & Moore, 1988a) implies that such reversion must be actively inhibited during development of the normal hymenium. Since explantation to media containing ammonium ions caused basidia, the only committed cells of the hymenium, to abort sporulation and revert to hyphal growth, normal sporulation requires protection from the inhibitory effects of metabolic sources of ammonium and related metabolites. Significantly, the ammonium assimilating enzyme NADP-dependent glutamate dehydrogenase (NADP-GDH) is derepressed specifically in basidia (Moore, 1984), being localised in microvesicles associated with the cell periphery (Elhiti, Moore & Butler, 1987) where it could serve as a detoxifying ammonium scavenger. Such a function might also be ascribed to the glutamine synthetase which is derepressed co-ordinately with NADP-GDH (Moore, 1984; Moore *et al.*, 1987).

Attempts to identify other specific inhibitory molecules on the basis of comparison of effects in other organisms have met with varied success. Ammonium depletion triggers a *Dictyostelium* slug to transform into a fruit body (Schindler & Sussman, 1977) and ammonium acts as an inhibitor of stalk-cell differentiation (Gross *et al.*, 1983). Such inhibitory effect was antagonised by 1 to 2 μM diethylstilboestrol and zearalenone, which inhibit plasma membrane ATPase (Gross *et al.*, 1983). Therefore, zearalenone and diethylstilboestrol were tested by transplanting gill segments of *Coprinus* to medium containing ammonium plus diethylstilboestrol (concentration range 1 to 800 μM) or zearalenone (100 to 600 μM). No antagonism between the additives was observed, implying that *Coprinus* and *Dictyostelium* do not share the same mechanism of action (Moore *et al.*, 1994).

The inhibition shown by ammonium may act through effects on the ionic balance of the basidium. The ammonium ion has the same dimensions as potassium, and its salts show many chemical similarities with salts of this metal. The effects of salts of potassium, sodium, rubidium, caesium, calcium and magnesium were assessed over the concentration range 25 to 150 mM (Chiu & Moore, 1990b). Exposure to alkali metal salts at any time from meiosis through to late sporulation stages resulted in premature termination of basidium development and outgrowth of vegetative hyphae. These results would not support the contention that the ammonium effect is solely due to this ion acting as an analogue of potassium. If this were so, one would expect the same spectrum of effects to be caused by rubidium but this was not observed. Clearly though, the ionic environment, as affected by both cations and anions, does seem to exert some influence.

As uptake of ammonium and other ions will affect the electrochemical gradient, the effects of the membrane-depolarizing agents dinitrophenol (DNP) and sodium azide (NaN_3), the sodium channel gramicidin S and the

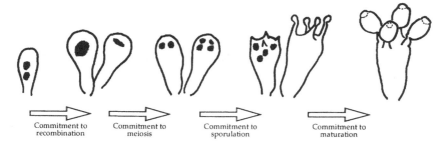

Fig. 3. The basidial differentiation pathway in *Coprinus cinereus*.

potassium carrier valinomycin were examined. Tissues excised when in meiotic stages were inhibited by exposure to sublethal 0.75 mM DNP or NaN$_3$; and whilst many basidia completed sporulation and the explanted gills autolysed, some showed reversion with hyphae growing out in place of sterigmata. Much the same result was obtained with gramicidin-S and valinomycin. With all ionophores, gills excised at sporulation stages were not affected but those excised during meiosis were sensitive and in each case an increased concentration of the agent increased the proportion of basidia affected. Since dinitrophenol and other membrane-depolarizing agents can raise the cAMP level in *C. cinereus* (Uno & Ishikawa, 1981) and other fungi (Trevillyan & Pall, 1979; Thevelein *et al.*, 1987), and cAMP levels during meiosis and sporulation were kept low in *C. cinereus* (Kuhad *et al.*, 1987), the effect of exposure of explanted gills to cAMP was also tested, and experiments were also carried out with tunicamycin and nikkomycin, which are potent inhibitors of wall synthesis (in *S. cerevisiae*, tunicamycin inhibited epispore formation (Weinstock & Ballou, 1987)). Like the ionophores, cAMP and wall synthesis inhibitors were effective only if applied during meiosis.

In brief, the differential sensitivity of basidia between meiotic and sporulation stages towards these diverse groups of inhibitors implies that during the nuclear division the cell is prepared in advance for sporulation, rather like the egg cell of an animal or plant, so that by the end of the cytologically recognizable nuclear division sporulation can proceed despite treatment with ionophores and wall synthesis inhibitors. Basidial differentiation in *Coprinus cinereus*, therefore, can be seen as a sequence of integrated steps of commitment (Fig. 3) consisting of the following landmarks: (i) commitment to recombination (requires completion of DNA synthesis; Lu, 1982); (ii) commitment to meiosis (at prophase I; Lu & Chiu, 1978; Chiu &

Moore, 1988a); (iii) commitment to sporulation (at or after meiotic II division; Raudaskoski & Lu, 1980; Chiu & Moore, 1988a & b); and (iv) commitment to maturation (Chiu & Moore, 1988b). It is not surprising to find a similar sequence of events in the meiocyte pathway of *S. cerevisiae* (Esposito & Klapholz, 1981).

In view of the wide range of compounds which have been tested in these bioassays, it is remarkable that the pattern of reversion was highly similar. Most hyphal apices were formed at sites expected to be involved in active wall synthesis during the normal progress of development. The effect of ammonium is unique: inhibiting sporulation and spore maturation *in vitro*. When the tissue exposed to ammonium treatment was in post-meiotic and early sporulation stages, the reversion hyphae grew out at the sites of sterigma formation; if the basidia had formed sterigmata, hyphae, instead of basidiospores, grew from their apices; if spores were in process of formation, exposure to ammonium caused termination of spore formation and direct outgrowth of hyphal tips from the immature spores still borne on basidia. It seems likely that ammonium ions interfere with cell wall metabolism and the cytoskeletal architecture which defines the number, position and nature of the outgrowths from the basidium.

Fungal walls and fruit body size

The fungal cell wall is a highly dynamic structure which is subject to modification during growth and development. Comparisons of wall composition between monokaryons and dikaryons of *C. cinereus* (Marchant, 1978; Blayney & Marchant, 1977) showed that whilst two compatible monokaryotic strains contained 33.4 and 26.8% (w/w) chitin, the chitin content of the walls of vegetative hyphae of their joint dikaryon was only 10.1%, although fruit body stems from the same dikaryon had a chitin content of 43.3%. Although there have been many studies of the architecture of fungal walls, particularly of the tip region (see Chapter 2), little is known of the dynamics, regulation and spatial organisation of wall synthesis and assembly.

The primary structural components of the fungal cell wall are polysaccharides which may be homo- or heteropolymers. Proteins are also significant components of the wall and are frequently covalently bonded to the polysaccharide constituents. Lipids and melanins are minor wall components in many fungi. Wall polysaccharides can be divided into two groups on the basis of their function and physical form. Structural (or skeletal) polysaccharides are water insoluble homopolymers and include chitin and β-linked glucans. The matrix polysaccharides are amorphous or slightly crystalline and are generally water soluble. Fungal walls are layered. The structural components (mainly chitin fibrils) are normally located on the inner side, frequently embedded in amorphous matrix material. The inner surface is characteristically fibrillar in appearance and is covered by protein and a glycoprotein reticulum with an outer layer of α- and β-glucans (Hunsley & Burnett, 1970; Cabib, Roberts & Bowers, 1982; Sietsma & Wessels, 1990). The cell wall of *C. cinereus* is composed of chitin, glucans with α-(1,4), β-(1,3) and β-(1,6) linkages, and glycans/glycoproteins containing xylomannans (Schaeffer, 1977; Kamada & Takemaru, 1977, 1983; Marchant, 1978; Bottom & Siehr, 1979, 1980).

Chitin microfibrils in stem cells of *C. cinereus* have been described as existing in right- or left-handed helices (Kamada *et al.*, 1991). About two-thirds are left handed, the sense of helicity is constant throughout a cell. The helicity was found to be the same before and after rapid elongation indicating that new microfibrils are inserted between existing ones. The onset of the helical arrangement of chitin microfibrils in the fruit body of *C. cinereus* was traced back to hyphal knots of 0.1 to 0.2 mm in diameter, prior to differentiation of recognisable fruit body tissues (Kamada & Tsuru, 1993). The walls of these hyphal knots contained lower amounts of chitin than hyphae in other developmental phases, so it appears that hyphal walls become differentiated in the fruit body before any obvious change in cell morphology. Mol, Vermeulen & Wessels (1990) found that in elongating hyphae of the mushroom stem of *Agaricus bisporus*, glucosaminoglycan chains were transversely oriented and not organised into distinct chitin microfibrils. Walls from the vegetative hyphae in the substrate showed randomly oriented microfibrils embedded in an amorphous matrix. A model was put forward explaining wall growth as occurring by diffuse extension. This was thought to be due to creeping of polymers in the walls caused by continuous breakdown and reformation of hydrogen bonds among the glucan chains and passive re-orientation of the glucosaminoglycan chains in a transverse manner.

Bartnicki-Garcia (1973) proposed a model for wall synthesis in which wall-bound lytic enzymes at the hyphal apex produce a controlled lysis to allow insertion of new polymers (see Chapter 2). It has also been proposed that wall-bound enzymes have a role in activating new growth points during branch formation. Synthesis of chitin is catalysed by the enzyme chitin synthase which catalyses transfer of *N*-acetylglucosamine from uridine diphosphate-*N*-acetylglucosamine to a growing chitin chain. Chitin synthases are integral membrane bound proteins and require solubilisation with digitonin before purification. The intracellular distribution and location of chitin synthase still provokes controversy. Although Sentandreu, Mormeneo & Ruiz-Herrera (1994) indicated that chitin synthesis could only occur inside the cell, it remains to be proven whether chitin synthase is located on the inner or outer side of the membrane (Duran, Bowers & Cabib, 1975; Vermeulen & Wessels, 1983; Kang et al., 1984).

Delivery of chitin synthetic machinery to the membrane seems to be the responsibility of chitosomes, which are microvesicles rich in zymogenic chitin synthase (Bartnicki-Garcia, Ruiz-Herrera & Bracker, 1979; Bartnicki-Garcia & Bracker, 1984). Chitosomes are intracellular microvesicles typically of 40–70 nm diameter. When activated in vitro by proteolytic enzymes and incubated in the presence of substrate UDP-GlcNAc, chitosomes generate chitin fibrils. One microfibril of chitin is produced from one vesicle. Although Cabib (1987) questioned whether chitosomes are real structures or whether they arise from disruption of other organelles it now seems generally accepted that at least '. . . some of the microvesicles observed at the apex of fungal hyphae may be chitosomes . . .' (Gooday, 1983). The current concept is that when vesicles meet the membrane at the hyphal tip they release their contents and merge with the wall.

In Coprinus cinereus, attempts have been made to isolate mutants resistant to calcofluor white. Calcofluor white is a fluor (commercially, a fabric whitening agent) which binds to nascent chitin microfibrils through hydrogen bonding with free hydroxyl groups and microfibril assembly by polymerisation is seriously disrupted. Calcofluor white-induced chitin products are profoundly different from the native microfibrillar one and abnormal walls are formed in yeasts exposed to the fluor (Elorza, Rico & Sentandreu, 1983; Roncero et al., 1988).

The experiment used $A_{mut}B_{mut}$ strains of C. cinereus which are homokaryotic phenocopies of the dikaryon (see below). A total of 1.5×10^6 oidia was used to select mutants showing resistance to hyphal growth inhibition by calcofluor white. Of the 60 calcofluor white (cfw) resistant strains isolated, 46 produced micro-fruit bodies (mean height 18 ± 6.4 mm) and 14 produced normal fruit bodies (mean height 45.2 ± 7.1 mm compared with the 48.2 ± 4.6 mm tall fruit bodies formed by the parental $A_{mut}B_{mut}$ strain).

True dikaryons made between cfw resistant strains and the parental strain (verified by their failure to produce oidia) were sensitive to growth inhibition by calcofluor white. This indicates that resistance to calcofluor white was a recessive character. All the cfw × cfw dikaryons were resistant to concentrations inhibiting the growth of the parental strain, suggesting all

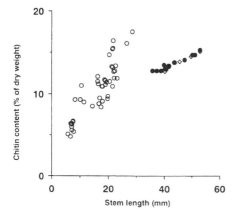

Fig. 4. Chitin content (% dry weight) compared with stem length (mm). The different symbols show data for micro-fruit body *cfw* resistant strains (open circles), macro-fruit body *cfw* resistant strains (closed circles) and the parental $A_{mut}B_{mut}$ strain (open diamonds). Means of 3 replicates.

were alleles of a single gene (*cfw*). Normal-sized *cfw* × micro-fruit body *cfw* dikaryons produced normal-sized fruit bodies, indicating that the micro-fruit body phenotype was also recessive, but micro-fruit body *cfw* × micro-fruit body *cfw* dikaryons only formed micro-fruit bodies. Thus, the apparently allelic *cfw* mutants were differentiated into two groups on the basis of fruit body size. Assay of chitin content in the cell walls showed that micro-fruit body *cfw* resistant strains contained significantly less chitin than normal (Fig. 4); mean values ($\pm$ SEM) being $10.4 \pm 0.5\%$ (of dry weight) for micro-fruit body *cfw* resistant strains, $13.6 \pm 0.2\%$ for macro-fruit body *cfw* resistant strains and $13.9 \pm 0.3\%$ for the parental $A_{mut}B_{mut}$ strain. Among the micro-fruit body strains, a positive correlation was observed between chitin content and stem length (Fig. 4). Interestingly, a negative correlation was observed between chitin content and gravitropic reaction time (Fig. 5). Horizontally placed stems of micro-fruit body *cfw* resistant strains responded to a change in the gravity vector within 57 ± 3.8 minutes, macro-fruit body *cfw* resistant strains respond within 42.7 ± 0.7 minutes compared with 40.2 ± 0.4 minutes for the parental strain. At the moment the mechanism(s) which relate(s) wall structure to fruit body size remain unknown.

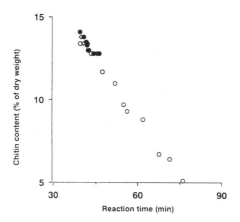

Fig. 5. Chitin content (% dry weight) compared with gravitropic reaction time. The different symbols show data for micro-fruit body *cfw* resistant strains (open circles), macro-fruit body *cfw* resistant strains (closed circles) and the parental $A_{mut}B_{mut}$ strain (open diamonds). Means of 3 replicates.

Using developmental mutants and spontaneous variants

Population polymorphisms and spontaneous mutants

Sclerotia of *Coprinus cinereus* are globose, small (*ca* 250 μm diam.) but multicellular, persistent resting structures. They are formed by vegetative mycelia, both monokaryotic and dikaryotic. Mushroom fruit bodies are borne more commonly on the dikaryon but monokaryotic fruiters are frequent. Both sclerotia and fruit bodies develop from undifferentiated mycelia through an organized process of hyphal growth and branching which forms an aggregate in which cellular differentiation occurs. The first study of the internal structure of mature sclerotia (Volz & Niederpruem, 1970) revealed that mature sclerotia possessed an outer unicellular rind layer, composed of cells with thickened and pigmented walls, which enclosed a medulla composed of a compact mass of thin-walled bulbous cells and accompanying hyphae. A later study (Waters, Butler & Moore, 1972, 1975a) reported a very different structure; the sclerotia were found to have a multilayered rind enclosing a compact medulla composed predominantly of thick-walled cells. Further analysis of sclerotia produced by 47 monokaryotic strains of different geographical origins showed that this difference in structure was a genuine polymorphism within the wild population of the species (Hereward & Moore, 1979).

The genetic basis of the polymorphism was investigated by constructing dikaryons from selected monokaryotic strains, the sclerotia produced being scored as having either a multilayered or single layered rind by light microscopy of wax-embedded sections. The monokaryons used for dikaryon construction included a number known to be unable to produce sclerotia because of genetic defects. Three loci (*scl-1, scl-2* and *scl-3*) at which alleles which prevent sclerotium formation have been identified were mapped genetically by Waters *et al.* (1975b) and representatives of these were combined in dikaryons together with strains able to produce either one of the two different sorts of sclerotium. The main conclusions were (i) the single layered rind was the dominant phenotype; (ii) the inability to form sclerotia was recessive; (iii) the different *scl*-negative genes complemented one another in doubly-heterozygous dikaryons (i.e. dikaryons made between sclerotium-negative strains were able to produce sclerotia).

It was concluded that the type of sclerotium with a multi-layered rind structure was the phenotype resulting from a genetic defect which was an allele of the gene *scl-1* (designated *scl-1*^{H}) which was recessive to wild-type but dominant to its silent allele (*scl-1*^{0} strains exhibiting the sclerotium-negative phenotype). The *scl-1*^{0} alleles segregated in crosses as expected of a single chromosomal gene (Waters *et al.*, 1975b) and were assumed to represent a gene responsible for some factor involved in sclerotium morphogenesis. A change in gene being able either to completely block sclerotium production (the *scl-1*^{0} phenotype) or to formation of sclerotia which were over endowed' with their component tissues (the *scl-1*^{H} phenotype). The *scl-1* gene product may thus be concerned with determining the extent and type of growth made during maturation of sclerotium initials.

Early stages of the pathway which leads to the formation of the fruit body (Matthews & Niederpruem, 1972) were strikingly similar to events described for the initiation of sclerotia (Waters *et al.*, 1975b). For this and other circumstantial reasons it was suggested that the two structures were alternative outcomes of a single initiating pathway in the dikaryon (Moore & Jirjis, 1976). A test of this proposition was performed using the sclerotium-negative (*scl*) strains to make homoallelic dikaryons (i.e. dikaryons in which both nuclei carried the same *scl* allele)(Moore, 1981). Of the four *scl* genes characterised; one, *scl-4*, caused abortion of developing fruit body primordia even when paired in the dikaryon with a wild type nucleus but the other three behaved as recessive genes in such heteroallelic dikaryons (and were mapped to existing linkage groups by Waters *et al.*, 1975b). Dikaryons which were homoallelic for any of the four *scl* genes were unable to form either sclerotia or fruit bodies. These observations con-

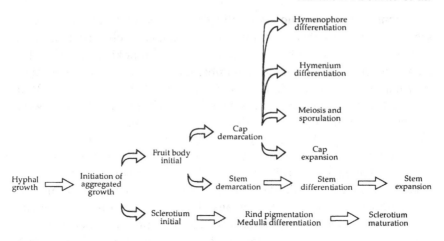

Fig. 6. The common initiation pathway to both fruit body and sclerotium development in *Coprinus cinereus*.

firmed earlier suspicions that a common pathway of initiation is used by both fruit bodies and sclerotia (Fig. 6).

Selection of developmental mutants

'Dominant' dikaryotic strains

The most extensive study of developmental mutants in mushrooms has been done with the Japanese strains of *C. cinereus*. It is worth emphasising that although these are called *C. macrorhizus* they have been demonstrated to be conspecific with the European isolates by mating tests (Moore *et al.*, 1979). Takemaru & Kamada (1972) isolated over 1,500 developmental variants following mutagen treatment of macerated dikaryon fragments. Among the mutants isolated were some (designated knotless') which were unable to differentiate. This phenotype was similar to that of homoallelic *scl*-negative dikaryons. However, since Takemaru & Kamada (1972) dealt solely with dikaryotic isolates it seems likely that their knotless mutants were dominant and thus different in nature from the recessive *scl*-genes. Penetrance of *scl*-genes in heteroallelic dikaryons depended on segregation of modifiers (Moore, 1981). The effect of these modifiers may help in understanding the very high frequency of dominant mutations observed by Takemaru & Kamada (1972). The assumption made by these authors was that their developmental variants arose as the result of mutations in genes controlling development. However, mutations in modifiers could allow both the expression of previously recessive variants present in the original genome, and also make more likely the expression of newly-induced developmental gene mutations. About 15% of the survivors of mutagen treatment were found to carry dominant developmental variations (Takemaru & Kamada, 1972), but nearly 75% of the variants were assigned (in about equal number) to just two phenotypes. It is possible that recessive genes for these phenotypes occurred in the parental dikaryon and that the high frequency of their occurrence in dominant form in the mutagen survivors was due to increased penetrance because of mutations in members of what could be a large and heterogeneous population of modifying loci, rather than to mutations in genes involved in morphogenesis.

These mutants were classified into categories on the basis of the phenotype of the fruit body produced: (i) 'knotless' - no hyphal aggregations formed; (ii) 'primordiumless' – aggregations were formed but did not develop further; (iii) 'maturationless' - primordia failed to mature; (iv) 'elongationless' – stem failed to elongate but cap development was normal (v) 'expansionless' - stem elongation normal but cap failed to open; (vi) 'sporeless' - few or no spores formed in what were otherwise a normal fruit bodies (Takemaru & Kamada, 1972). These mutant phenotypes suggest that different aspects of fruit body development are genetically separate, and fruit body morphogenesis is under polygenic control. Prevention of meiosis still permits the fruit body to develop normally. However, the hymenia of both the 'elongationless' and 'expansionless' mutants showed frequent abnormal basidia (2- and 3-spored basidia and basidia with asynchronously maturing basidiospores) (Chiu, unpublished observations). Therefore, the defective development of the fruit body may create different microhabitats in the hymenophore leading to abnormality in sporulation. Perhaps more interesting is the fact that mutants were obtained with defects in either cap expansion or stem elongation. Both processes depend on enormous cell inflation (Gooday, 1985; Moore *et al.*, 1979), and the fact that they can be separated by mutation indicates that the same result (increase in cell volume) is achieved by different means.

Haploid fruiters especially with double mutations in the mating type factors

Normally in *C. cinereus*, basidiospores germinate to form mycelium which has cells containing a single haploid nucleus. Mating between these monokaryons is controlled by an outbreeding mechanism dependent on the operation of two incompatibility (or mating type) factors termed A and B each of which harbours a series of linked genes of multiple alleles (Casselton *et al.*, 1995). For an attempted mating to form a dikaryon, the two parental monokaryons must carry different alleles at both A and B loci, such a pairing is said to be compatible. The A and B genes are located on different chromosomes and segregate at random during meiosis. In a successful mating, hyphae from monokaryons of compatible mating types anastomose, forming a dikaryon which has two nuclei (one of each mating type) within each cell. This mycelium grows forming clamp connections at each newly formed septum, one of the two nuclei divides here while the other divides in the main part of the cell under the regulation of the mating type genes (Casselton *et al.*, 1995). Upon suitable environmental trigger and the presence of functional fruit body developmental genes, a dikaryon gives rise to a fruit body and meiosis occurs in basidia in the hymenium of the gills of the fruit body cap, so completing the life cycle.

The fact that cells in the tissues of the fruit body contain two (parental) nuclei means that each cell is effectively a genetic diploid and presents the inevitable problem for genetic analysis of development that recessive mutations will not be expressed. Swamy *et al.* (1984) described a strain of C. *cinereus* carrying mutations in both of its mating type factors ($A_{mut}B_{mut}$). This strain is a homokaryotic dikaryon phenocopy. It is similar to the dikaryon in that the hyphae have binucleate compartments and extend by conjugate nuclear division with formation of clamp connections, and the cultures can produce apparently normal fruit bodies. On the other hand, it is a homokaryon, being able to produce asexual spores (usually called oidia) and, most importantly, containing only one (haploid) genetic complement.

Kanda & Ishikawa (1986) and later followed by Zolan, Tremel & Pukkila (1988) showed how such strains could be used for the isolation and characterization of developmental mutants (Kanda *et al.*, 1989a & b; 1990). These strains have especially been used to study meiosis and spore formation (Zolan *et al.* 1988; Kamada *et al.*, 1989; Kanda *et al.*, 1989a, 1990; Pukkila *et al.*, 1995; Zolan *et al.*, 1995; see Chapter 5). The sporeless (or, better, sporulation-deficiency; e.g. Fig. 7) mutants obtained so far showed defects in either one of the four steps of commitment identified in the meiocyte pathway (Fig. 2). They are grouped as: (i) failure to complete premeiotic DNA replication leading to arrestment at meta-anaphase I; (ii) meiotic mutants with defects in DNA repair machinery leading to formation of defective synaptonemal complex in meiosis and hypersensitivity towards radiation; (iii) failure to initiate sporulation; and (iv) failure to

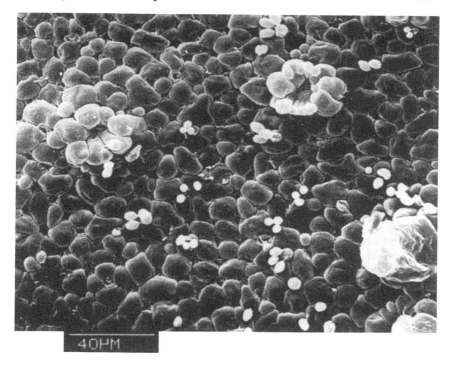

Fig. 7. Scanning electron micrograph showing the hymenial phenotype of a leaky radiation-sensitive mutant in *Coprinus cinereus*. Most of the basidia are arrested at prophase I and do not sporulate while some complete meiosis but, abnormally, bear 2 to 3 basidiospores.

produce mature basidiospores (Miyake, Takemaru & Ishikawa, 1980; Miyake, Tanaka & Ishikawa, 1980; Pukkila *et al.*, 1995; Zolan *et al.*, 1995).

The mutant revoluta

The *revoluta* strain was isolated from a parental strain of genotype $A_{mut}B_{mut}$, *paba*-1 after ultraviolet irradiation; *revoluta* is a morphological mutant producing abnormal fruit bodies. Heterokaryotic dikaryons resulting from mating with other monokaryons all produced normal fruit bodies, so *revoluta* is recessive and it segregates in a simple Mendelian pattern. Characteristically, the *revoluta* fruit body had a convoluted hymenophore with revolute cap margin and short solid stem (Fig. 8). The first gills to be formed were radial but as the hymenophore grew, the gills became increasingly convoluted and the margin of the cap became revolute.

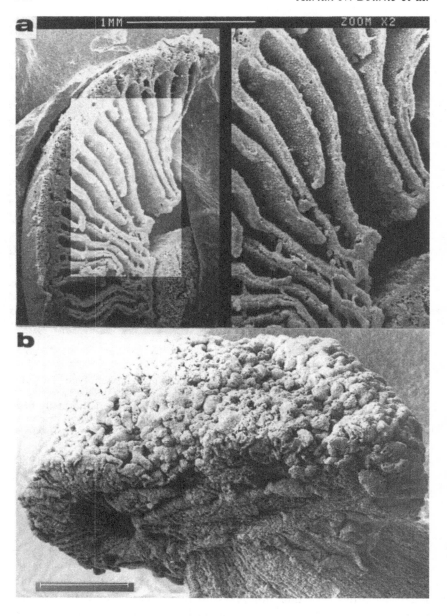

Fig. 8. Scanning electron micrographs showing the hymenophores of wild type (a) and strain *revoluta* (b) of *Coprinus cinereus*. a, the horizontal section shows the radially arranged thin gills around the central stem; b, the longitudinal half fruit body shows the convoluted hymenophore. Bar = 200 μm.

The *revoluta* mutation was clearly pleiotropic in that stem development, cap morphology, hymenophore initiation and developmental synchrony all differed from normal. But the crucial abnormality was that the mutant lacked the primary gill connection to the stem at its very early appearance of the hymenophore. As a result of the lack of gill anchorage and the uneven distribution of the cystidia-cystesia pairs between neighbouring gills, stresses generated during cap growth could not be properly directed. Consequently, the 'embryonic' convoluted hymenophore present in normal fruit bodies persisted instead of the normal radially symmetrical gills around the central stem at maturity (Chiu & Moore, 1990a). Basidia did sporulate on this convoluted hymenophore but the normally-observed synchrony was lost; adjacent gills could be at different stages of sporulation. Therefore, the enclosed cap with the primary gills connected to the central stem, a characteristic feature of genus *Coprinus*, is essential for maintaining synchrony of meiosis in basidia, another unique feature of the genus.

Conclusions

Experiments described here highlight a number of experimental approaches which could be more widely applied in other organisms to examine pattern formation in mushrooms, such as continuous video observation, explantation, use of inhibitors or metabolite analogues and genetic mutants or natural variants. In other chapters of this book, we show the use of chemical techniques, statistical analysis and video-imaging to give us a clear picture of fruiting morphogenesis in *C. cinereus*. The study of the *revoluta* strain and related investigations of spontaneous developmental polymorphisms in *Volvariella bombycina* have suggested that normal fruit body development comprises a sequence of independent but coordinated morphogenetic subroutines, each of which can be activated or repressed as a complete entity (Chiu, Moore & Chang, 1989; see Chapters 1 & 8). In this model the 'hymenium subroutine' (for example) in an agaric would be invoked normally to form the 'epidermal' layer of the gill; a 'hymenophore subroutine' producing the classic agaric gill plates. In any one species the subroutines are invoked in a specific sequence which generates the particular ontogeny and morphology of that species. Abnormal fruit bodies which arise spontaneously are interpreted as being produced by correct execution of a morphogenetic subroutine which has been invoked in the wrong place or at the wrong time. In a morphologically different species the subroutines are invoked in yet a different sequence, or with different timing. The model provides a unifying theme for categorising fruit body ontogeny and for clarifying phylogenetic and taxonomic relationships (Watling & Moore, 1993; and see Chapters 1 & 8).

References

Anderson, G. E. (1971). *The Life History and Genetics of* Coprinus lagopus. Philip Harris Biological Supplies Ltd.: Weston-super-Mare, U.K.

Bartnicki-Garcia, S. (1973). Fundamental aspects of hyphal morphogenesis. In *Microbial Differentiation*, Symposium of the Society for General Microbiology, **vol. 23** (ed. J. O. Ashworth & J. E. Smith), pp. 245–267. Cambridge University Press: Cambridge, U.K.

Bartnicki-Garcia, S. & Bracker, C. E. (1984). Unique properties of chitosomes. In *Microbial Cell Wall Synthesis and Autolysis* (ed. C. Nombela), pp. 101–112. Elsevier: Amsterdam.

Bartnicki-Garcia, S., Ruiz-Herrera, J. & Bracker, C. E. (1979). Chitosomes, and chitin synthesis. In *Fungal Walls and Hyphal Growth* (ed. J. H. Burnett & A. P. J. Trinci), pp. 149–168. Cambridge University Press: Cambridge, U.K.

Bastouill-Descollonges, Y. & Manachère, G. (1984). Photosporogenesis of *Coprinus congregatus*: correlations between the physiological age of lamellae and the development of their potential for renewed fruiting. *Physiologia Plantarum* **61**, 607–610.

Berry, D. R. (1983). Ascospore formation in yeast. In *Fungal Differentiation: a Contemporary Synthesis* (ed. J. E. Smith), pp. 147–173. Marcel Dekker: New York.

Blayney, G. P. & Marchant, R. (1977). Glycogen and protein inclusions in elongating stipes of *Coprinus cinereus. Journal of General Microbiology* **98**, 467–476.

Bottom, C. B. & Siehr, D. J. (1979). Structure of an alkali-soluble polysaccharide from the hyphal wall of the basidiomycete *Coprinus macrorhizus* var. *microsporus. Carbohydrate Research* **77**, 169–181.

Bottom, C. B. & Siehr, D. J. (1980). Structure and composition of the alkali-insoluble cell wall fraction of *Coprinus macrorhizus* var. *microsporus. Canadian Journal of Biochemistry* **58**, 147–153.

Brunt, I. C. & Moore, D. (1989). Intracellular glycogen stimulates fruiting in *Coprinus cinereus. Mycological Research* **93**, 543–546.

Cabib, E. (1987). The synthesis and degradation of chitin. *Advances in Enzymology* **59**, 59–101.

Cabib, E., Roberts, R. & Bowers, B. (1982). Synthesis of the yeast cell wall and its regulation. *Annals of Biochemistry* **52**, 763–793.

Casselton, L. A., Asante-Owusu, R. N., Banham, A. H., Kingsnorth, C. S., Kes, U., O'Shea, S. F. & Pardo, E. H. (1995). Mating type control of sexual development in *Coprinus cinereus. Canadian Journal of Botany* **73**, S266–S272.

Chiu, S. W. & Moore, D. (1988a). Evidence for developmental commitment in the differentiating fruit body of *Coprinus cinereus. Transactions of the British Mycological Society* **90**, 247–253.

Chiu, S. W. & Moore, D. (1988b). Ammonium ions and glutamine inhibit sporulation of *Coprinus cinereus* basidia assayed *in vitro. Cell Biology International Reports* **12**, 519–526.

Chiu, S. W. & Moore, D. (1990a). A mechanism for gill pattern formation in

Coprinus cinereus. Mycological Research **94**, 320–326.

Chiu, S. W. & Moore, D. (1990b). Sporulation in *Coprinus cinereus*: use of an *in vitro* assay to establish the major landmarks in differentiation. *Mycological Research* **94**, 249–253.

Chiu, S. W. & Moore, D. (1993). Cell form, function and lineage in the hymenia of *Coprinus cinereus* and *Volvariella bombycina*. *Mycological Research* **97**, 221–226.

Chiu, S. W., Moore, D. & Chang, S. T. (1989). Basidiome polymorphism in *Volvariella bombycina*. *Mycological Research* **92**, 69–77.

Chiu, S. W. & To, S. W. (1993). Endogenous glycogen is not a trigger for fruiting in *Pleurotus sajor-caju*. *Mycological Research* **97**, 363–366.

Dawes, I. W. (1983). Genetic control and gene expression during meiosis and sporulation in *Saccharomyces cerevisiae*. In *Yeast Genetics: Fundamental and Applied Aspects* (ed. J. F. T. Spencer, D. M. Spencer & A. R. W. Smith), pp. 29–64. Springer-Verlag: New York, Heidelberg, Berlin.

Delavier-Klutchko, C., Durian-Trautmann, O., Allemad, P. & Tarlitzki, J. (1980). Assimilation of ammonia during sporogenesis of *Saccharomyces cerevisiae. Journal of General Microbiology* **116**, 143–148.

Duran, A., Bowers, B. & Cabib, E. (1975). Chitin synthase zymogen is attached to the yeast plasma membrane. *Proceedings of the National Academy of Sciences, U.S.A.* **72**, 3952–3955.

Elhiti, M. M. Y., Moore, D. & Butler, R. D. (1987). Ultrastructural distribution of glutamate dehydrogenases during fruit body development in *Coprinus cinereus. New Phytologist* **107**, 531–539.

Elorza, M. V., Rico, H. & Sentandreu, R. (1983). Calcofluor white alters the assembly of chitin fibrils in *Saccharomyces cerevisiae* and *Candida albicans* cells. *Journal of General Microbiology* **131**, 2209–2216.

Esposito, R. E. & Klapholz, S. (1981). Meiosis and ascospore development. In *The Molecular Biology of the Yeast Saccharomyces cerevisiae,* **Vol. 1,** *Life Cycle and Inheritance* (ed. J. N. Strathern, E. W. Jones & J. R. Brach), pp. 211–287. Cold Spring Harbor Laboratory: New York.

Ewaze, J. O., Moore, D. & Stewart, G. R. (1978). Co-ordinate regulation of enzymes involved in ornithine metabolism and its relation to sporophore morphogenesis in *Coprinus cinereus. Journal of General Microbiology* **107**, 343–357.

Freese, E. B., Olempske-Beer, Z., Hartig, A. & Freese, E. (1984). Initiation of meiosis and sporulation of *Saccharomyces cerevisiae. Developmental Biology* **102**, 438–451.

Gooday, G. W. (1983). The hyphal tip. In *Fungal Differentiation: A Contemporary Synthesis* (ed. J. E. Smith), pp. 315–356. Marcel Dekker: New York.

Gooday, G. W. (1985). Elongation of the stipe of *Coprinus cinereus*. In *Developmental Biology of Higher Fungi* (ed. D. Moore, L. A. Casselton, D. A. Wood & J. C. Frankland), pp. 311–331. Cambridge University Press: Cambridge, U.K.

Gross, J. D., Bradbury, J., Kay, R. R. & Peacey, M. J. (1983). Intracellular pH and the control of cell differentiation in *Dictyostelium discoideum. Nature* **303**, 244–245.

Hammad, F., Ji, J., Watling, R. & Moore, D. (1993). Cell population dynamics in *Coprinus cinereus*: co-ordination of cell inflation throughout the maturing basidiome. *Mycological Research* **97**, 269–274.

Hereward, F. V. & Moore, D. (1979). Polymorphic variation in the structure of aerial sclerotia of *Coprinus cinereus*. *Journal of General Microbiology* **113**, 13-18.

Hunsley, D. & Burnett, J. H. (1970). The ultrastructural architecture of the walls of some hyphal fungi. *Journal of General Microbiology* **62**, 203-216.

Ji, J. & Moore, D. (1993). Glycogen metabolism in relation to fruit body maturation in *Coprinus cinereus*. *Mycological Research* **97**, 283-289.

Jirjis, R. I. & Moore, D. (1976). Involvement of glycogen in morphogenesis in *Coprinus lagopus*. *Journal of General Microbiology* **95**, 348-352.

Kamada, T., Kurita, R. & Takemaru, T. (1978). Effects of light on basidiocarp maturation in *Coprinus macrorhizus*. *Plant & Cell Physiology* **19**, 263-275.

Kamada, T., Sumiyoshi, T., Shindo, Y. & Takemaru, T. (1989). Isolation and genetic analysis of resistant mutants to the benzimidazole fungicide benomyl in *Coprinus cinereus*. *Current Microbiology* **18**, 215-218.

Kamada, T. & Takemaru, T. (1977). Stipe elongation during basidiocarp maturation in *Coprinus macrorhizus*: changes in polysaccharide composition of stipe cell wall during elongation. *Plant & Cell Physiology* **18**, 1291-1300.

Kamada, T. & Takemaru, T. (1983). Modifications of cell wall polysaccharides during stipe elongation in the basidiomycete *Coprinus cinereus*. *Journal of General Microbiology* **129**, 703-709.

Kamada, T., Takemaru, T., Prosser, J. I. & Gooday, G. W. (1991). Right and left handed helicity of chitin microfibrils in stipe cells of *Coprinus cinereus*. *Protoplasma* **165**, 65-70.

Kamada, T. & Tsuru, M. (1993). The onset of the helical arrangement of chitin microfibrils in fruit body development of *Coprinus cinereus*. *Mycological Research* **97**, 884-888.

Kanda, T., Arakawa, H., Yasuda, Y. & Takemaru, T. (1990). Basidiospore formation in a mutant of the incompatibility factors and mutants that arrest at meta-anaphase I in *Coprinus cinereus*. *Experimental Mycology* **14**, 218-226.

Kanda, T., Goto, A., Sawa, K., Arakawa, H., Yasuda, Y. & Takemaru, T. (1989a). Isolation and characterization of recessive sporeless mutants in the basidiomycete *Coprinus cinereus*. *Molecular and General Genetics* **216**, 526-529.

Kanda, T., Ishihara, H. & Takemaru, T. (1989b). Genetic analysis of recessive primordiumless mutants in the basidiomycete *Coprinus cinereus*. *Botanical Magazine, Tokyo* **102**, 561-564.

Kanda, T. & Ishikawa, T. (1986). Isolation of recessive developmental mutants in *Coprinus cinereus*. *Journal of General and Applied Microbiology* **32**, 541-543.

Kang, M. S., Elango, N., Mattia, E., Au-Young, J., Robbins, P. W. & Cabib, E. (1984). Isolation of chitin synthase from *Saccharomyces cerevisiae*. Purification of the enzyme by entrapment in the reaction products. *Journal of Biological Chemistry* **259**, 14966-14972.

Kuhad, R. C., Rosin, I. V. & Moore, D. (1987). A possible relation between cyclic-AMP levels and glycogen mobilisation in *Coprinus cinereus*. *Transactions of the British Mycological Society* **88**, 229-236.

Lu, B. C. (1982). Replication of deoxyribonucleic acid and crossing over in *Coprinus*. In *Basidium and Basidiocarp: Evolution, Cytology, Function and Development* (ed. K. Wells & E. K. Wells), pp. 93-112. Springer-Verlag:

New York, Heidelberg & Berlin.

Lu, B. C. & Chiu, S. M. (1978). Meiosis in *Coprinus*. IX. The influence of premeiotic S-phase arrest and cold temperature on the meiotic cell cycle. *Journal of Cell Science* **32**, 21–30.

Madelin, M. F. (1960). Visible changes in the vegetative mycelium of *Coprinus lagopus* Fr. at the time of fruiting. *Transactions of the British Mycological Society* **43**, 105–110.

Marchant, R. (1978). Wall composition of monokaryons and dikaryons of *Coprinus cinereus*. *Journal of General Microbiology* **106**, 195–199.

Matthews, T. R. & Niederpruem, D. J. (1972). Differentiation in *Coprinus lagopus*. I. Control of fruiting and cytology of initial events. *Archives of Microbiology* **87**, 257–268.

McLaughlin, D. J. (1982). Ultrastructure and cytochemistry of basidial and basidiospore development. In *Basidium and Basidiocarp: Evolution, Cytology, Function and Development* (ed. K. Wells & E. K. Wells), pp. 37–74. Springer-Verlag: New York, Heidelberg, Berlin.

Miller, J. J. (1963). Determination by ammonium of the manner of yeast nuclear division. Nature **198**, 214.

Miyake, H., Takemaru, T. & Ishikawa, T. (1980). Sequential production of enzymes and basidiospore formation in fruiting bodies of *Coprinus macrorhizus*. *Archives of Microbiology* **126**, 201–205.

Miyake, H., Tanaka, K. & Ishikawa, T. (1980). Basidiospore formation in monokaryotic fruiting bodies of a mutant strain of *Coprinus macrorhizus*. *Archives of Microbiology* **126**, 207–212.

Montgomery, G. P. & Lu, B. C. (1990). Involvement of *Coprinus* endonuclease in preparing substrate for *in vitro* recombination. *Genome* 33, 101–108.

Mol, P. C., Vermeulen, C. A. & Wessels, J. G. H. (1990). Diffuse extension of hyphae in stipes of *Agaricus bisporus* may be based on a unique wall structure. *Mycological Research* **94**, 480–488.

Moore, D. (1981). Developmental genetics of *Coprinus cinereus*: genetic evidence that carpophores and sclerotia share a common pathway of initiation. *Current Genetics* **3**, 145–150.

Moore, D. (1984). Developmental biology of the *Coprinus cinereus* carpophore: metabolic regulation in relation to cap morphogenesis. *Experimental Mycology* **8**, 283-297.

Moore, D. (1987). The formation of agaric gills. *Transactions of the British Mycological Society* **89**, 105-108.

Moore, D., Elhiti, M. M. Y. & Butler, R. D. (1979). Morphogenesis of the carpophore of *Coprinus cinereus*. *New Phytologist* **83**, 695-722.

Moore, D., Hammad, F. & Ji, J. (1994). The stage in sporulation between the end of meiosis and emergence of sterigmata is most sensitive to ammonium inhibition in *Coprinus cinereus*. *Microbios* **76**, 197–201.

Moore, D., Horner, J. & Liu, M. (1987). Co-ordinate control of ammonium-scavenging enzymes in the fruit body cap of *Coprinus cinereus* avoids inhibition of sporulation by ammonium. *FEMS Microbiology Letters* **44**, 239-242.

Moore, D. & Jirjis, R. I. (1976). Regulation of sclerotium production by primary metabolites in *Coprinus cinereus* (= *C. lagopus sensu* Lewis). *Transactions of the British Mycological Society* **66**, 377-382.

Moore, D., Liu, M. & Kuhad, R. C. (1987). Karyogamy-dependent enzyme derepression in the basidiomycete *Coprinus*. *Cell Biology International*

Reports **11**, 119 - 124.

Moore, D. & Pukkila, P. J. (1985). *Coprinus cinereus*: an ideal organism for studies of genetics and developmental biology. *Journal of Biological Education* **19**, 31–40.

Morimoto, N. & Oda, Y. (1973). Effects of light on fruit-body formation in a basidiomycete, *Coprinus cinereus*. *Plant & Cell Physiology* **14**, 217–225.

Morimoto, N., Suda, S. & Sagara, N. (1981). Effect of ammonia on fruit-body induction of *Coprinus cinereus* in darkness. *Plant & Cell Physiology* **22**, 247–254.

Piñon, R. (1977). Effects of ammonium ions on sporulation of *Saccharomyces cerevisiae*. *Experimental Cell Research* **105**, 367–378.

Pukkila, P. J., Shannon, K. B. & Skrzynia, C. (1995). Independent synaptic behavior of sister chromatids in *Coprinus cinereus*. *Canadian Journal of Botany* **73**, S215–S220.

Raju, N. B. & Lu, B. C. (1973). Meiosis in *Coprinus*. IV. Morphology and behavior of spindle pole bodies. *Journal of Cell Science* **12**, 131-141.

Raudaskoski, M. & Lu, B. C. (1980). The effect of hydroxyurea on meiosis and genetic recombination in the fungus *Coprinus lagopus*. *Canadian Journal of Genetics and Cytology* **22**, 41-50.

Roncero, C., Valdivieso, M. H., Ribas, J. C. & Duran, A. (1988). Isolation and characterisation of *S. cerevisiae* mutants resistant to calcofluor white. *Journal of Bacteriology* **170**, 1945–1949.

Rosin, I. V. & Moore, D. (1985). Origin of the hymenophore and establishment of major tissue domains during fruit body development in *Coprinus cinereus*. *Transactions of the British Mycological Society* **84**, 609-619.

Schaeffer, H. P. (1977). An alkali-soluble polysaccharide from the cell walls of *Coprinus lagopus*. *Archives of Microbiology* **113**, 79–82.

Schindler J. & Sussman M. (1977). Ammonia determines the choice of morphogenetic pathways in *Dictyostelium discoideum*. *Journal of Molecular Biology* **116**, 161–169.

Sentandreu, R., Mormeneo, S. & Ruiz-Herrera, J. (1994). Biogenesis of the fungal cell wall. In *The Mycota*, **Vol. I**, *Growth, Differentiation and Sexuality* (ed. J. G. H. Wessels & F. Meinhardt), pp. 111–124. Springer-Verlag: Berlin.

Sietsma, J. H. & Wessels, J. G. H. (1990). The occurrence of glucosaminoglycan in the wall of *Schizosaccharomyces pombe*. *Journal of General Microbiology* **136**, 2261–2265.

Slack, J. M. W. (1983). *From Egg to Embryo: Determinative Events in Early Development*. Cambridge University Press: Cambridge, U.K.

Swamy, S., Uno, I & Ishikawa, T. (1984). Morphogenetic effects of mutations at the *A* and *B* incompatibility factors in *Coprinus cinereus*. *Journal of General Microbiology* **130**, 3219–3224.

Swamy, S., Uno, I. & Ishikawa, T. (1985). Regulation of cyclic AMP metabolism by the incompatibility factors in *Coprinus cinereus*. *Journal of General Microbiology* **131**, 3211–3217.

Takemaru, T. & Kamada, T. (1972). Basidiocarp development in *Coprinus macrorhizus*. I. Induction of developmental variations. *Botanical Magazine* (*Tokyo*) **85**, 51-57.

Thevelein, J. M., Beullens, M., Honshoven, F., Hoebeeck, G., Detremerie, K., Den Hollander, J. A. & Jans, A. W. H. (1987). Regulation of the cAMP level in the yeast *Saccharomyces cerevisiae*: intracellular pH and the effect of membrane depolarizing compounds. *Journal of General Microbiology* **133**, 2191-2106.

Trevillyan, J. M. & Pall, M. L. (1979). Control of cAMP levels by depolarizing agents in fungi. *Journal of Bacteriology* **138**, 397-403.

Uno, I. & Ishikawa, T. (1981). Control of adenosine 3',5'-monophosphate level and protein phosphorylation by depolarizing agents in *Coprinus macrorhizus. Biochimica et Biophysica Acta* **672**, 108-113.

Uno, I. & Ishikawa, T. (1982). Biochemical and genetical studies on the initial events of fruitbody formation. In *Basidium and Basidiocarp: Evolution, Cytology, Function and Development* (ed. K. Wells & E. K. Wells), pp. 113–123. Springer-Verlag: New York, Heidelberg & Berlin.

Vermeulen, C. A. & Wessels, J. G. H. (1983). Evidence for a phospholipid requirement of chitin synthase in *Schizophyllum commune. Current Microbiology* **8**, 67–71.

Volz, P. A. & Niederpruem, D. J. (1970). The sclerotia of *Coprinus lagopus. Archiv für Mikrobiologie* **70**, 369–377.

Waters, H., Butler, R. D. & Moore, D. (1972). Thick-walled sclerotial medullary cells in *Coprinus lagopus. Transactions of the British Mycological Society* **59**, 167-169.

Waters, H., Butler, R. D. & Moore, D. (1975a). Structure of aerial and submerged sclerotia of *Coprinus lagopus. New Phytologist* **74**, 199-205.

Waters, H., Moore, D. & Butler, R. D. (1975b). Morphogenesis of aerial sclerotia of *Coprinus lagopus. New Phytologist* **74**, 207-213.

Watling, R. & Moore, D. (1994). Moulding moulds into mushrooms: shape and form in the higher fungi. In *Shape and Form in Plants and Fungi* (ed. D. S. Ingram & A. Hudson), pp. 270–290. Academic Press: London.

Weinstock, K. G. & Ballou, C. E. (1987). Tunicamycin inhibition of epispore formation in *Saccharomyces cerevisiae. Journal of Bacteriology* **169**, 4384-4387.

Zolan, M. E., Tremel, C. J. & Pukkila, P. J. (1988). Production and characterization of radiation-sensitive meiotic mutants of *Coprinus cinereus. Genetics* **120**, 379-387.

Zolan, M. E., Stassen, N. Y., Ramesh, M. A., Lu, B. C. & Valentino, G. (1995). Meiotic mutants and DNA repair genes of *Coprinus cinereus. Canadian Journal of Botany* **31**, S226-S233.

Chapter 7

Control of growth and patterning in the fungal fruiting structure. A case for the involvement of hormones

LILYANN NOVAK FRAZER

Summary

Development and tissue differentiation in animal and plant multicellular organisms results from communication between cells via hormones and growth factors. The patterns associated with tissue differentiation and growth in fungal fruiting structures have been observed and described at the macroscopic, cellular, ultrastructural and even genetic/molecular levels. Although the formation of a fully differentiated mushroom, composed of distinct, organised tissues, from an initial mass of uncoordinated hyphae, provides the *prima facie* evidence for fungal hormones or morphogens coordinating differentiation and growth in developing fruit bodies, this has not been substantiated since few chemical candidates have been identified. This is in stark contrast to the animal and plant kingdoms where hormones and growth factors have been discovered, identified, purified and are commercially available. Fungal hormones are known to occur in lower fungi (oomycetes, chytridiomycetes, zygomycetes) where they are essential for mating, sexual development and differentiation and there is some evidence for their presence during differentiation of sexual fruiting structures in higher fungi (ascomycetes, basidiomycetes) but there is little information about hormonal coordination of growth during the development of fruit bodies or vegetative, multihyphal structures. The possibility that hormones or growth factors may be involved in the development of sexual and vegetative fruiting structures and in differential growth during tropic responses to external stimuli is discussed.

Introduction

Development in multicellular organisms is complex and depends on a multitude of morphological, physiological and genetic changes regulated in part by morphogens, hormones and growth factors. In animals, morphogens are important in setting up patterns in the mass of unspecialised cells comprising the embryo, resulting in the development of a mature, fully differentiated organism. Hormones are important in regulating the physiology of the organism, from the embryonic to the mature adult stage, and growth factors regulate directed and differential growth. Peptide and steroid hormones have long been known to regulate the physiology of organisms including man (Nicola, 1994) while morphogens, such as retinoic acid, active in the development of limb buds in chick embryos (Brickell & Tickle, 1989; Tickle, 1991) and the recently discovered protein products of the *bicoid*, *dorsal* and *hedgehog* genes in *Drosophila* (Ip, Levine & Small, 1992; Lawrence, 1992; St Johnston & Nüsslein-Volhard, 1992), are important in establishing the positional information required for cell differentiation at the embryonic stage.

While retinoic acid could be considered a 'classical' morphogen in that it controls gene expression indirectly by binding to plasmalemma and nuclear membrane receptors (Tickle, 1991) via a concentration gradient, the *bicoid*, *dorsal* and *hedgehog* gene products, transcription factors which bind to specific DNA sequences and either stimulate or repress gene expression directly, also function in a concentration-dependent manner (Ip *et al.*, 1992) and thus can also be considered morphogens. Similarly, in plants, auxins are responsible for controlling cell division and differentiation at all stages of plant development (Sachs, 1991), cytokinins and gibberellins promote overall growth while abscissic acid and other compounds inhibit growth (Salisbury & Ross, 1985). Auxins and other plant growth regulators have also been found in fungi but these compounds have never been shown to have a hormonal role in the fungi that produce them or in other fungi (reviewed by Gruen, 1982). Their occurrence in phytopathogenic fungi is thought to be a consequence of secondary metabolism or a result of uptake from the host (Gooday, 1994). Remembering that gibberellins were named for the fungus from which these plant hormones were isolated, the prime role of any fungus-produced plant hormone is most likely to be to modify the growth of plant tissues which the fungus parasitises.

Several mating hormones have been characterised in a few fungal species and their role in sexual reproduction has been reviewed recently (Dyer, Ingram & Johnstone, 1992; Gooday & Adams, 1992; Bölker & Kahmann, 1993; Duntze, Betz & Nientiedt, 1994; Gooday, 1994; Mullins, 1994; Staben, 1995). On the other hand, evidence of hormonal involvement in coordinating other developmental processes, such as primordium induction and fruit body differentiation and maturation, is sparse, while belief in their involvement in tropic responses and even in the formation of multihyphal, vegetative structures is based on indirect and mainly circumstantial evidence.

The underlying rationale for investigating the presence and possible roles of morphogens, hormones and growth factors in fungi is based on the perceived level of evolutionary conservation in fundamental processes among diverse organisms. For example, transcription factors show a remarkable diversity in sequence variation and yet the homology of DNA-binding sites among taxonomically unlike species is striking as in the cases described above: the *bicoid* and *hedgehog* proteins contain homeobox motifs while the *dorsal* protein contains a REL domain related to that found in mammalian regulatory factor NF-kB and in the vertebrate oncoprotein *rel* (Ip *et al.*, 1992). Tymon *et al.* (1992) determined that the *A* factor mating type in *Coprinus cinereus* encoded a transcription factor, responsible for regulating both sexual and asexual development, which contains a POU domain similar in part to the bipartite DNA binding domain in certain animal transcription factors. The evidence that transcription factors, with DNA-binding sites sharing homology with those found in animals and plants, are also present in fungi (Kelly *et al.*, 1988; Kües *et al.*, 1992) begs the question of how widespread their presence and function is in multihyphal, fungal structures.

In this chapter, the possibility of morphogen, hormone and growth factor control at all stages of multihyphal development, both sexual and vegetative, is discussed on the basis of information derived from morphological, physiological and genetic/molecular biological analyses of development in a variety of species, predominantly filamentous basidiomycetes. Special emphasis is placed on *Coprinus cinereus*, which was the model organism used in experiments to be discussed in the final section (and see Chapters 1 and 6).

Fungal sex hormones, mating type gene products and fruiting inducers

The evidence for the presence and function of hormones (specifically, pheromones) is well-established in lower fungi (chytridiomycetes, oomycetes, zygomycetes) where they are involved in coordinating sexual differentiation and mating (see reviews by Gooday & Adams, 1992; Gooday, 1994; Mullins, 1994). In higher fungi, mating through the activity of hormones is well known in ascomycetous and basidiomycetous yeasts (Kelly *et al.*, 1988; Dyer *et al.*, 1992; Bölker & Kahmann, 1993; Duntze *et al.*, 1994) and there is some evidence that diffusible factors also have a role in inducing ascogonial and trichogyne formation in some filamentous ascomycetes (reviewed in Dyer *et al.*, 1992). In contrast, anastomosis between monokaryons (even incompatible ones) occurs freely in filamentous basidiomycetes and cell fusion between compatible monokaryons is thought to occur without the need for specific mating hormones (Bölker & Kahmann, 1993). But recent analysis of the mating type factor B in *Schizophyllum commune* has shown that its sequence shares homology with the pheromones and pheromone receptors of lower fungi (Kües & Casselton, 1992), thus a cell signalling mechanism seems to be required for effective mating even in this filamentous basidiomycete.

Dikaryon formation in filamentous basidiomycetes is under the control of mating type loci and the decision to undergo sexual development after cell fusion is initiated and regulated by interactions of regulatory proteins (putative transcription factors) related to the homeodomain proteins of higher eukaryotes (Kelly *et al.*, 1988; Banuett, 1992; Kües & Casselton, 1992, 1994; Kües *et al.*, 1992, 1994; Wessels, 1993a; Casselton & Kües, 1994; Glass & Nelson, 1994; Kämper, Bölker & Kahmann, 1994; Wessels, 1992, 1994; Staben, 1995). These mating type loci are usually thought of, and described, as the master regulatory genes for sexual development but present knowledge limits their activity to the initial steps in the process (Kües & Casselton, 1992), in particular mating resulting in dikaryon formation and regulation of nuclear segregation through formation of clamp connections. Whether products of these gene complexes control any steps in fruit body formation or maturation is unknown.

The dikaryotic phase in filamentous ascomycetes is limited to the ascogenous hyphae, any fruiting structures produced are constructed from the heterokaryotic hyphae. In most basidiomycetes (but see Chapter 5) the mating process forms a dikaryon which can grow vegetatively for an indeterminate period before sexual reproduction is induced by external factors, such as light, aeration and nutrient depletion (Wessels, 1993a; 1994). These external stimuli presumably activate expression of fruiting genes which eventually enable formation of a fruit body primordium. The basidiomycete primordium is normally comprised of heterokaryotic hyphae but only the basidia embark on karyogamy, meiosis and sporulation (see Chapters 5 & 6). Cap and stem tissues differentiate early in the initial stages of primordium formation (Chapter 1), grow, expand and mature into a fruit body (Fig. 1).

Probably the most fundamental change which occurs after sensitisation by the external factors mentioned above is the mechanism by which heterokaryotic hyphae, growing in an outwardly, diffuse, vegetative man-

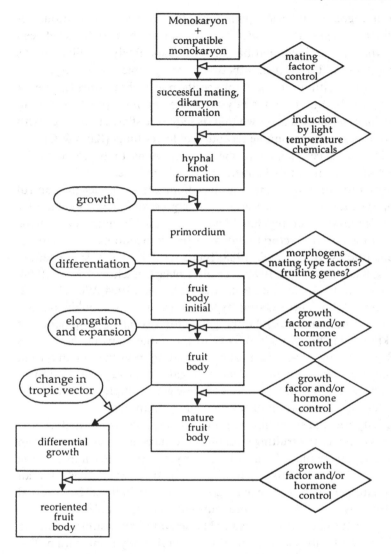

Fig 1. Potential target sites for growth factors and/or hormones during mating, dikaryon formation, primordium induction, fruit body differentiation and maturation.

ner, and yet primed for sexual development, are stimulated to grow together, to branch, and to co-operate in formation of the fruit body initial as a hyphal knot which is the building block for more intricate tissues within the fruit body (Moore, 1995). Unfortunately, we are completely ignorant of how this morphogenetic change is brought about. Light is the primary trigger for fruiting induction in many basidiomycetes (Ross, 1985; Kozak & Ross, 1991; Wessels, 1993a; 1994) and acts on plasma membrane receptors to sensitise hyphal tips for sexual development. How hyphal tips find each other and then have their branching pattern changed in a co-ordinated way is unknown.

Homing reactions have been described (e.g. Kemp, 1977; and see discussion in Moore, 1984b) in which hyphal tips tend to grow towards germinating spores of the same species. However, this is, by definition, a reaction between different individuals of the same species; its mechanism is unknown, as is its relevance to co-operation between hyphae of the same individual to produce a fruiting structure. Bringing hyphal tips together and co-ordination of branching patterns are obvious targets for hormonal control but very few chemicals which may function in these ways during the inductive process have been identified (reviewed by Uno & Ishikawa (1982) and Wessels (1993a)): cAMP and AMP in *C. cinereus* (Uno & Ishikawa, 1971, 1973a & b, 1982), cerebrosides in *Schizophyllum commune* (Kawai & Ikeda, 1982), an unidentified low molecular weight compound from *Agaricus bisporus* causing fruiting induction in *S. commune* (Rusmin & Leonard, 1978) and an unidentified, diffusible factor(s) in *Phellinus contiguus* (Butler, 1995) have been found to induce mono- and dikaryotic fruiting but the manner in which these substances interact with the fungus to elicit fruiting is unknown. The low molecular weight fruiting inducing substance extracted from *A. bisporus* is interesting because its cross-reactivity on the taxonomically unrelated *Schizophyllum commune* and its general chemical similarity to substances with growth factor-like properties extracted from *Flammulina velutipes* (Table 1) and *C. cinereus* (see below) imply that there may be a family of chemically related factors or hormones responsible for mediating primordium initiation and fruit body elongation. The recent discovery of hydrophobins, hydrophobic proteins abundantly expressed and deposited in the cell walls during emergent growth in *Schizophyllum commune* (Wessels, 1993a, 1994), reveals another class of molecules which may be involved in hyphal aggregation by manipulating the biophysics of the surface of the aggregate.

Detection of fruit body inducing substances is difficult as their production is, presumably, limited to very low concentrations as they are likely to be secreted into the medium in order to attract tips together over a very restricted area. The complex media which are usually needed to induce fruiting also create difficulties, so it is not surprising that so few chemicals have been identified as putative inducers. Molecular techniques, as those

Table 1. *Growth factor-type substances, extracted from basidiomycetous fruit bodies, with fruiting inducing, stem elongation and/or cap expansion promoting properties*

Source organism	Activity	Active substance	Stability	Solubility	Authors
Agaricus bisporus Flammulina velutipes Lentinus edodes Pleurotus ostreatus	mycelial growth; fruiting induction	whole fruitbody extract, active principle unknown, but ninhydrin +ve, reducing sugar +ve, organic acid -ve, phosphate -ve, fatty acid -ve	heat, acid, base stable	water, aqueous methanol (insoluble in absolute methanol, chloroform, petroleum, benzene)	Urayama (1969)
Coprinus cinereus	fruiting induction	cAMP, 3'-AMP, theophylline		water	Uno & Ishikawa (1971, 1973a & b)
Coprinus cinereus	fruiting induction	whole culture extract, active principle unknown			Uno & Ishikawa (1971, 1973a & b)
Coprinus cinereus	hyphal aggregates (fruiting induction)	cAMP			Matthews & Niederpruem (1972)
Agaricus bisporus	fruiting induction (in Schizophyllum commune)	fruitbody tissue extract, active principle unknown but <12,000 molecular weight	heat, acid, base stable	water, 50% ethanol, 50% acetone	Rusmin & Leonard (1978)
Schizophyllum commune	fruiting induction	cerebrosides (glycosphingolipid)	unknown	acetone	Kawai & Ikeda (1982)
Phelinus contiguus	fruiting induction	mycelial extract, active principle unknown	heat stable	water	Butler (1995)

Species	Effect	Active principle / notes	Stability	Solubility	Reference
Agaricus bisporus	stem elongation, cap expansion	gills, active principle unknown	unknown	none	Hagimoto & Konishi (1959)
Agaricus bisporus *Coprinus macrorhizus* = *cinereus*?) *Armillaria matsutake* *Hypholoma fasciculare*	stem elongation, cap expansion	gills, active principle unknown. *(contains IAA, but inactive)	heat, acid, base stable	ether, acetone, ethanol, water (insoluble in petroleum ether, benzene)	Hagimoto & Konishi (1960)
Agaricus bisporus	stem elongation; converts tryptophan to IAA	gills, active principle unknown but <12,000 molecular weight. *(contains IAA, but inactive)	unknown	none	Konishi & Hagimoto (1961)
Agaricus bisporus	stem elongation	gills, active principle unknown. Glutamic acid, leucine, cysteine, glycine, serine, asparagine, glutamine, threonine, tyrosine, valine, proline, arginine, $(NH_4)_2SO_4$ & NH_4Cl all tested +ve in the bioassay	unknown	ether, acetone, ethanol (pet. ether, benzene insoluble)	Konishi (1967)
Agaricus bisporus	stem elongation	gills, active principle unknown.	unknown	none	Gruen (1963)
Flammulina velutipes	stem elongation	gills, active principle unknown.	unknown	none	Gruen (1969)
Coprinus radiatus	stem elongation	gills, active principle unknown.	unknown	none	Eilers (1974)
Flammulina velutipes	stem elongation	gills, active principle unknown.	unknown	none	Gruen (1976)

Table 1. (cont.)

Source organism	Activity	Active substance	Stability	Solubility	Authors
Flammulina velutipes	stem elongation	gills, active principle unknown, but <12,000 molecular weight	heat stable	none (diffused into agar blocks)	Gruen (1982)
Coprinus congregatus	stem elongation	cap inhibitor, active principle unknown, but <12,000 molecular weight	unknown	none	Robert & Bret (1987)
Coprinus congregatus	stem elongation	cap inhibitor and stimulator, active principle unknown	unknown	unknown	Robert (1990)
Agaricus bisporus	mycelial growth, stem elongation	wound hormone, ODA (= 10-oxo-trans-8-decenoic acid), enzymatic degradation product of linoleic acid	unknown	extraction patented	Mau et al. (1992)
Coprinus micaceous	unknown	*cytokinin-like activity in fruit bodies	unknown	unknown	Szabo et al. (1970) (cited by Gruen, 1982)
Lentinus tigrinus	unknown	*auxin-like, gibberellin-like, cytokinin-like activities in stems and caps	unknown	unknown	Rypacek & Sladky (1972, 1973) (cited by Gruen, 1982)
Agaricus bisporus Boletus elegans Grifola frondosa Phallus impudicus Phellinus pomaceous	unknown	*gibberellin-like activity in caps	unknown	unknown	Pegg (1973)

* indicates the presence of plant growth regulators which have no stem elongation or cap expansion properties on the fungal fruit bodies tested (see Gruen, 1982).

used previously by Yashar & Pukkila (1985) for *C. cinereus* and by Mulder & Wessels (1986) for *S. commune*, that is, analysing mRNA expression (but at low levels) at different but specific stages should be more productive in identifying hormones than trying to extract them from media.

Differentiation within the primordium

The initial hyphal knot comprises more or less similar, intertwining hyphae (prosenchyma). There are many studies describing the morphological, physiological and even genetic and molecular changes which are correlated with fruit body formation (reviews by Gooday, 1982; Moore, 1984a & b, 1995; Manachère, 1988; Reijnders & Moore, 1985; Rosin, Horner & Moore, 1985; Watling & Moore, 1994; Wessels, 1993a, 1994) but few studies indicate how these various processes might be coordinated between these two stages.

There is some evidence for the involvement of particular chemicals during sexual differentiation in filamentous ascomycetes (e.g. production of linoleic acid in *Ceratocystis* spp., *Neurospora crassa*, *Nectria haematococca* and mycosporines in *Pyronema omphalodes*, *Morchella esculenta*, *Nectria galligena* (reviewed by Dyer *et al.*, 1992)) but whether these compounds act as morphogens or metabolites is not established. The theories and experimental evidence correlating morphological, physiological and genetic changes with differentiation in sexual fruit bodies are covered in more detail in Chapter 1. Since there is no direct experimental evidence for the existence of morphogens in the differentiating primordium it is useful to draw from other biological systems where more information exists about the coordination of developmental processes. An animal analogy of the early primordial stage is the chick limb bud, a multicellular but undifferentiated structure, while a useful plant analogy is vascular differentiation in plant embryos.

In the chick limb bud, retinoic acid, a derivative of vitamin A, is an endogenous signalling substance which specifies the position of mesenchymal cells within the embryo based on the gradient of retinoic acid in the interstitial space around cells (Tickle, 1989, 1991). The activity of retinoic acid is dose- and position-dependent and signals the formation of digits in the wing. Retinoic acid also regulates the expression of several homeobox genes, which specify for the anterior position of digits in the limb bud, by binding to nuclear retinoic acid receptors (Tickle, 1991). Thus, retinoic acid acts as a morphogen (dose-dependent specification of position of cells) and regulates gene expression (homeobox genes); its action via a concentration gradient in the interstitial environment establishes the cell-to-cell communication required for the formation of digits in the correct orientation.

In the case of vascular differentiation in plant embryos, auxin transport is instrumental in establishing polarised differentiation of vessels and cell shape in plant embryos via a concentration gradient and even acts during later stages when regenerative processes are required, for example during wound repair (Sachs, 1991). Exogenously added auxin alters gene expression by enhancing rapid transcription of specific mRNAs in plant tissues undergoing either cell elongation or cell division (Key, 1989; Guilfoyle *et al.*, 1993) and there is also evidence that auxin is bound to nuclear receptors (Löbler & Klämbt, 1985; Guilfoyle *et al.*, 1993; Ulmasov *et al.*, 1995 and references therein). Auxin-regulated mRNAs show distinct patterns of organ-specific, tissue-specific and development-specific expression (Guilfoyle *et al.*, 1993).

In *Dictyostelium discoideum* (which produces an asexual fruiting structure), starved amoebae are triggered to aggregate together into a mound by chemotaxis relayed by cAMP signals (Kay, Berks & Traynor, 1989). Cells of the mound differentiate further, into stalk or spore cells, via a gradient of diffusible differentiation inducing factors or DIFs (a family of chlorinated alkyl phenones) (Kay, Berks & Traynor, 1989; Berks *et al.*, 1991).

As in the previous examples, DIFs induce specific gene expression via a cytosolic and nuclear binding protein (receptor) (Williams *et al.*, 1987; Insall & Kay, 1990). An important criterion in all three systems cited here as examples is that morphogenesis is polarised, that is, it proceeds in a certain direction. Polarisation involves controlling gene expression in a graded manner yet gene activity is not known to be directional (it is on/off) so polarised development must be expressed as a coordination of events of many cells (Sachs, 1991), i.e. external factors control gene expression within cells. Also, in the case of retinoic acid and auxin, the response to the morphogen is dependent on the tissue it acts upon and not specific to the morphogen molecule. In contrast, DIFs in *D. discoideum* cause the specific differentiation of immobilised amoebal cells into prestalk, and then stalk cells (Kay *et al.*, 1989).

Are there parallels between these examples and primordium differentiation? Although there are light microscopic observations of early primordium structure (Reijnders & Moore, 1985 and references therein), no studies have explored whether there is hormonal specification of, for example, cap and stem regions in the developing primordium. The early primordial stages of *S. commune* and *C. cinereus* were described by van der Valk & Marchant (1978) as consisting of randomly oriented hyphae; at a later stage, hyphae were enveloped in a mucilage and clear demarcations of an apical centre in *S. commune* and of distinct cap and stem zones in *C. cinereus* were observed. Reijnders & Moore (1985) described that the early organisation of a primordium consists of two types of tissue, a bundle of nearly parallel hyphae and interwoven hyphae which form the plectenchyma. Microscopic studies describing morphogenesis in *C. cinereus* indicate that differentiation into cap and stem regions occurs very early in the development of a fruit body initial (Moore, Elhiti & Butler, 1979; and see Chapter 1). Similarly, in *A. bisporus*, 2 mm tall primordia consist of a disoriented mass of hyphae (Flegg & Wood, 1985) and yet by the time it reaches 6–10 mm in diameter, the primordium has differentiated into the tissue zones present in mature mushrooms. Clearly, even at this early stage, there is a polarity within the primordium, one of the requirements for establishing an actual/active morphogen gradient.

It is difficult to envisage that the primary differentiation of a primordium into distinct zones of differentiated tissues is attributable solely to the initial mating event. The process cannot represent the playing out of a sequence initiated by the first interaction of the mating type genes. It is far easier to believe that these fundamental differentiation events are directed by morphogens as in the case of retinoic acid regulation of digit formation in the chick limb bud (Tickle, 1991) or auxin-regulated vascular differentiation in plant stems (Sach, 1991). However, if the mating type genes are the master regulators of development, potential candidates for the control could be the proteins encoded by these genes or genes under their control. But first it would have to be demonstrated that the putative protein products of these genes did have transcription factor function which was active in a concentration gradient-dependent manner. Neither of these conditions have yet been met.

An important feature of fungal development which is different from animal and, to a degree, plant systems is that most differentiated fungal tissue is not determined or irreversibly committed to a specific fate. Consequently, the differentiation which occurs is not terminal and all the tissues of a fruit body, other than the probasidia which are determined after prophase I (Chiu & Moore, 1988, 1990), can dedifferentiate and revert to vegetative growth upon transplantation to a new medium (see Chapter 6). Thus it is extremely important that a differentiating 'environment' is maintained within the intact tissue, possibly via morphogens, to ensure fulfilment of development. Primordia are often enveloped in a mucilage (van der Valk & Marchant, 1978) which could serve as the medium through which morphogens could maintain the differentiated state. Investigations with developing primordia, i.e. dissection and exogenous application of morphogens as was done to determine the presence and action of morphogens in the chick limb bud system (Brickell & Tickle, 1989), are difficult due to their small size and the unpredictability of the location of their formation; but the problems are not insurmountable (Chapter 6) and molecular approaches could also be useful. Although there are no reports in the literature of studies investigating the genetic changes correlated specifically with the formation of the cap or stem zones as yet, this field is worth further examination.

Growth and maturation of the fruit body

Fruit body growth involves cell expansion in species such as *C. cinereus* (Gooday, 1985; Hammad *et al.*, 1993; Hammad, Watling & Moore, 1993), cell expansion as well as division in other species, such as *Agaricus bisporus* (Craig, Gull & Wood, 1977) and only cell division in *S. commune* (Wessels, 1992). Different aspects of stem elongation during fruit body maturation have been reviewed by Gooday (1985) and recently by Kamada (1994). In plants and animals hormones and growth factors involved in morphogenesis as well as cell division and expansion are the subjects of active study. Sadly, this is not true for the fungi. With the one exception of 10-oxo-*trans*-8-decenoic acid (ODA, an enzymatic breakdown product of linoleic acid) (Tressl, Bahri & Engel, 1982), a hormone produced in *A. bisporus* as a result of wounding (Mau, Beelman & Ziegler, 1992), astonishingly little attention has been given to the compounds with hormone or growth factor-like activity described in earlier studies of extracts from *A. bisporus* (Hagimoto & Konishi, 1959, 1960; Konishi & Hagimoto, 1961; Gruen, 1963; Konishi, 1967), *F. velutipes* (Gruen, 1969, 1976) and other basidiomycetes (Hagimoto & Konishi, 1960; Eilers, 1974; Robert & Bret, 1987; Robert, 1990). The last extensive review of this topic appeared fourteen years ago (Gruen, 1982)!

Previous attempts to purify and identify fungal substances which might regulate fruit body growth are shown in Table 1. In most cases, the objective of these studies was simply to determine whether hormone-like or growth factor-like compounds existed in fungi. Presence or absence was the essential criterion and any active ingredients found in fruit body extracts were not chemically identified; only their capacity to promote stem elongation and/or cap expansion was described. Konishi (1967) was the first to partially purify a substance from *A. bisporus* caps which enhanced stem elongation in the *Agaricus* test and determine that the growth factor was comprised of various amino acids, which he tested (in pure solution) for their individual effects on stem elongation. Whether these amino acids were functioning as individual growth factors or as nutrients (at a concentration of 10^{-4} M) is not known and there seems to have been no effort to purify the active, fungal ingredient further.

Considering that the evidence for the presence of fungal growth factors is so fragmentary and derives from experiments involving very different species and diverse extraction methods, it is surprising that the various extracts (including those analyzed for their ability to elicit fruit body induction/formation) have exhibited similar activities and chemical properties. Extracts from *A. bisporus, Coprinus macrorhizus, Hypholoma fasciculare, Armillaria matsutake* (Hagimoto & Konishi, 1960; Konishi, 1967; Urayama, 1969), *Lentinus edodes, F. velutipes,* and *Pleurotus ostreatus* (Urayama, 1969; Gruen, 1982) all cause stem elongation and cap expansion, are 12,000 molecular weight, heat stable, acid/base stable and mostly soluble in polar solvents including water. These similarities in characteristics may suggest that the active compounds comprise a family of hormones or growth factors of slightly different chemical structure in each species, but with enough similarities to be cross-reactive, as in the case of the *A. bisporus* extracts which induce fruiting in *S. commune* (Rusmin & Leonard, 1978), and broadly similar in gross chemical character. This is not unlike the situation in plants where auxin is actually a family of related compounds based on indole-3-acetic acid (Salisbury & Ross, 1985) and active on a very wide variety of plants. While the majority of the substances in Table 1 stimulate extension, those isolated by Robert & Bret (1987) and Robert (1990) from *C. congregatus* have inhibitory activities and were extracted from fruit bodies at earlier stages of development (primordium). This is evidence that both inhibitory and stimulatory substances are produced in fruit bodies during growth and that there may be temporal control of growth factor activity or production which may be important at different stages of development. In contrast, the fruiting induction substance extracted by Rusmin & Leonard (1978) from different developmental stages of *A. bisporus* fruit bodies showed no differences in activities (that is, all stages produced equivalent fruiting inducing activity). Technically, these two cases are not totally comparable since the first pertains to a factor(s) which causes enhanced or inhibited stem extension while the second deals with fruiting induction.

Clearly, there is not yet enough evidence of the activity and identity of these substances to understand the manner in which such growth factor-like compounds regulate fruit body development. Purification is the key. Purification and determination of the complete chemical structure of the active ingredient. Many of the difficulties in purification attempts have been compounded by the use of complex agar media to isolate growth factor-type activity. Indiscriminate extraction processes, for example from whole fruit bodies, from damaged fruit bodies or using other than the gentlest extraction techniques, may have released cytosolic components able to degrade the activity of the very compounds the experimenters were trying to isolate. It would also be difficult to isolate a large enough quantity of growth factor if only small numbers of fruit bodies were used in the extraction processes, especially if the factors occur at concentrations lower than 10^{-6} M.

Probably the greatest area for error is the bioassay. Without a sensitive and appropriate bioassay, activity of the compounds in question may be missed altogether or conversely, the activity observed may not be solely due to hormonal effects but also due to nutrient metabolism. Most of the studies reported in the literature were performed over a very long time (24-72 h). Such lengthy assays immediately pose the question of whether stems, detached from the parent mycelium, were still viable especially in the case of *A. bisporus* and *F. velutipes* where full or normal extension of the fruit body is dependent on the parent mycelium (Gruen, 1982). An even more relevant question is whether the putative growth factor is likely to be active over the sort of time scale used for published bioassays. By definition, growth factors (especially those controlling small morphogenetic fields) must be unstable, either intrinsically or through active destruction, as one of the ways to establish the concentration gradient. Improvements to the bioassay techniques might include using species which react rapidly and/or assay criteria which can be judged more quickly so that results can be obtained in a few hours rather than days and by limiting the potential number of compounds being bioassayed by performing simple separation techniques, such as dialysis (Rusmin & Leonard, 1978; Gruen, 1982; Robert & Bret, 1987), gel filtration (Rusmin & Leonard, 1978) or thin-layer chromatography.

Vegetative structures

Multihyphal vegetative organs (mycelial strands, coremia, rhizomorphs and sclerotia) may also develop under the coordinating signal(s) of growth factors. These multihyphal structures are composed of distinct tissue zones, such as the rind, cortex and medulla of many sclerotia (Willetts & Bullock, 1992) and they perform an important role in the survival of the organism, be it nutrient retrieval in the case of coremia and rhizomorphs (Watkinson, 1979) or overwintering in the case of sclerotia (Willetts & Bullock, 1992;

Moore, 1995). Such structures are formed from vegetative, monokaryotic hyphae and so, mating type genes do not operate in this situation. Thus the same question needs to be posed about how the change comes about from a diffusely-growing, vegetative pattern of hyphal growth to one in which hyphae grow together in harmony as an aggregate, but without the involvement of any of the mechanisms which might be involved in finding and attracting mates.

The repulsion normally encountered by hyphae in the same colony must be replaced by attraction or, in the very least, by a neutral or no response between adjacent hyphae (Willetts & Bullock, 1992). These structures are highly differentiated; for example, sclerotia are composed of distinct tissues, each comprised of hyphae which differ in structure and chemical properties from the vegetative hyphae from which they originated (Willetts & Bullock, 1992). The hyphae comprising rhizomorphs, which are specialised for efficient translocation of nutrients and are therefore scavenging organs of phytopathogenic species such as *Armillaria mellea* (Watkinson, 1979), are physiologically different from the vegetative hyphae from which they developed. Thus, there is a need to determine whether chemical factors may be involved in coordinating the initial formation of these organised structures and once they are formed, whether there is a need for growth factor control between the different tissues comprising them. Many of these structures produce mucilage (Watkinson, 1979) or a layer of reactive quinone compounds (as a result of phenoloxidase activity which is elevated in vegetative, multihyphal structures (Willetts & Bullock, 1992)) and these extracellular matrices may facilitate intercellular communication via growth factors which can direct morphogenesis of hyphae in tissue zones in sclerotia and rhizomorphs. Again, hydrophobins may also be involved, perhaps in an adhesive role, since hydrophobins specific to monokaryons are produced (reviewed in Wessels, 1993b) and hydrophobic interactions between hyphae emerging from a medium may be sufficient to bind the hyphae together loosely.

Tropisms as morphogenetic changes

Fungi have evolved strategies for different tissues to develop and grow in different directions with respect to gravity, light and other external stimuli. For example, in mushrooms the cap expands revealing gills, pores, tubes or teeth on the underside, which must be oriented vertically downwards to allow the basidiospores to escape from the fruit body. The positive gravitropism of the gills and the negative gravitropism of the stem are the mechanisms which achieve the vertical orientation. In *Coprinus* spp. the gills are agravitropic so the vertical position of the cap is solely dependent on the responses of the stem (Moore, 1991). The tropic response involves the tissue somehow sensing the orientation stimulus (direction of 'down' for gravitropism, direction of 'brightest' for phototropism, etc.) and then differentially regulating the growth of its constituent cells so that the organ is repositioned. Thus a tropic response is a convenient tool for the study of morphogenesis since the application of the stimulus is in the hands of the experimenter and the response to the stimulus involves differential regulation of cell differentiation. We have done most work on gravitropism in *C. cinereus*.

When a mushroom is laid horizontally it is able to reorient its hymenium to the correct position by the stem growing differentially, raising the cap to the correct orientation (reviewed in Moore, 1991; Moore *et al.*, 1996). Thus a change in the gravity vector induces a simple, developmental pattern-forming process, that is a morphogenetic change, whereby the perceived external signal (change in the gravity vector) is transduced into a biological/cellular response resulting in regeneration of the hymenium or controlled differential growth. The gravitropism of *C. cinereus* has been used as a model to generate this morphogenetic change on demand and to look for hormonal or growth factor control of elongation during the differential growth generated for the gravitropic response.

The kinetics of the gravitropic response in the two most studied basidiomycetes, *C. cinereus* and *F. velutipes*, have recently been compared and reviewed (Moore *et al.*, 1996) and reveal evidence for the presence of growth factors controlling differential growth. In *C. cinereus*, the initial response is due to the lower hyphae (meaning those hyphae in the lower half of a horizontal stem) elongating faster than the upper hyphae (meaning those hyphae in the upper half of the same horizontal stem). Light microscopic studies reveal that lower hyphae increase in length by 4–5 fold without increase in girth, and growth studies indicate that the lower surface of the stem elongates at a faster rate than the upper surface to generate the gravitropic bend (Greening & Moore, 1996). Differential extension is achieved without an increase in the number of lower hyphae or inflation of narrow hyphae in the lower half or conversely, a decrease in the number of hyphae or deflation of inflated hyphae in the upper half (Greening & Moore, 1996). Clearly, the fact that different regions of the same gravitropically responding stem extend at different rates implies not only that the relative position of hyphae in the stem is recognised but that there must be a mechanism by which differential growth is coordinated.

These results, along with those presented below, implicate growth factor control in one of three ways: (i) a gradient of a stimulatory substance is established which induces extension of the lower hyphae preferentially; (ii) a gradient of an inhibitory substance is established which inhibits extension of the upper hyphae preferentially; or (iii) both types of substance are present/produced but their distribution in the stem results in the extension pattern described above. Curvature begins at the apex of the stem and proceeds in a basipetal direction (Kher *et al.*, 1992); it does not proceed in an apical direction nor remain at the location in which the bend originates. This implies that the transport of growth factor(s) is basipetal. Once the responding stem has reached a certain curvature (estimated to be approximately 35° (Moore *et al.*, 1994)), the apex begins to straighten out or unbend so that it does not overshoot the vertical. This phenomenon is a separate process called curvature compensation (Kher *et al.*, 1992) and it implies a change in the transport or in the balance of growth signals present or active in the responding stem.

When a horizontal stem is attached by the apex rather than the base it initially responds normally, by lifting the basal portion of the stem until it reaches the vertical but it then continues to bend and curls over, often into a circle (Kher *et al.*, 1992). These observations indicate that the apex must be free to move for curvature compensation to occur, imply that there is polar transport of the compensation signal (from the apex to the base) and that there is interaction of growth signals during the normal response controlling the different extension rates of lower and upper portions of the stem. Preliminary experiments have been performed to determine whether stem hyphae react individually or in a concerted manner during the gravitropic response. Stems were split longitudinally (with the base left intact) and placed horizontally in a moist chamber (as described in Kher *et al.*, 1992); each piece exhibited negative gravitropism. Even though some pieces curled back in opposite directions, because of the release of tissue tensions when first cut, each portion always returned to the vertical.

Intact gravitropically stimulated stems still respond and grow normally under water (Kher *et al.*, 1992), but when longitudinally split stems were placed under water, they did not uncurl and did not return to the vertical even though extension was unaffected. This indicates that water immersion dissipated the polar transport or gradient of growth factors (Novak Frazer, unpublished results) and also shows that the growth factors were active in the extracellular matrix of the stem. Experiments have also demonstrated that the stem is polarised with regard to the distribution of gravireceptive cells. There are fewer perceptive cells at the base than at the apex, so a faster response is realised at the apex (Moore *et al.*, 1994). The gradient of receptive cells (many at the apex, fewer at the base) has been shown by sequential removal of sections of stem and monitoring response of the remaining part (Greening, Holden & Moore, 1993): stem response was delayed in direct relationship to the amount of apex is removed. These results, although circumstantial, indicate the potential for growth factor involvement during gravimorphogenesis in *C. cinereus*. The ingredients for growth factor/hormonal control are present: the stem is a polar system and different regions within the stem (that is, the lower and upper portions) respond in a different manner, both physiologically and temporally.

Potential growth factors have been extracted from pre- and post-meiotic fruit bodies and two different activities found, possibly corresponding to two different substances (Moore & Novak Frazer, 1993, 1994). The bioassay used was similar to that described by Michalenko (1971) and Gruen (1982) although lanolin was not used as a carrier. One substance (Fungiflex 1) was produced by both immature and mature fruit bodies of *Coprinus cinereus* but the other (Fungiflex 2) was only produced by mature fruit bodies. Fungiflex 2 was found to be between 10–100 times more abundant in the cap than in the stem. Both substances were of low molecular weight ($>12,000$ molecular weight), as determined by dialysis, gel filtration and mass spectroscopy, and were heat stable. In these two respects, molecular weight and heat stability, the substances extracted are similar to those described in earlier studies (Table 1). Fungiflex 1 has a faster action (results seen after 1 h exposure) and has inhibitory properties, similar to substances extracted from *C. congregatus* (Bret & Robert, 1987). Fungiflex 2 has a slower action (results seen after 6 h exposure) and has stimulatory properties, similar to substances extracted from various basidiomycetes (see Gruen, 1982 for review).

Notwithstanding the previous similarities, Fungiflex 1 and 2 function in a much shorter time frame than those already described (bioassay results visible over 1–18 h rather than the 24–72 h time frame of other studies). This may be partially explained by the fact that *C. cinereus* is a relatively fast-growing and probably fast-reacting species compared to *Agaricus* and *Flammulina*. Fungiflex 1 and 2 are similar in their solubility characteristics (water and methanol soluble) to those of other extracted growth factors although generally soluble in fewer solvents (Table 1). The identity of Fungiflex 1 and 2 is unknown as they have not yet been purified.

Although these substances may function during the gravitropic response to generate differential growth and were originally extracted from gravireacting stems, it is unlikely that they are specifically gravitropic hormones or growth factors, whose synthesis is induced by the external stimulus. The fact that Fungiflex 1 and 2 (or substances with the same activities as those extracted from gravitropically reacting stems) can also be extracted from pre- and post-meiotic and vertically growing fruit bodies indicates that they may have a more general role in coordinating growth. Their similarity to other fungal growth factors (Table 1) may reflect a general role in controlling extension. During rapid elongation, which occurs soon after meiosis in *C. cinereus* (Hammad *et al.*, 1993), the growth of the expanding cap and elongating stem must be coordinated. The expansion of the whole fruit body must also proceed accurately in a vertical direction so that the gill surface is properly oriented when sporulation occurs a few hours later. Thus these growth factors are probably involved in continually correcting the direction of growth of the mushroom as well as coordinating the expansion of the cap and elongation of the stem during the later stages of fruit body development.

Future studies

Bioassays are cumbersome to perform, and if performed with impure extracts, which by their nature are variable, are sometimes impossible to interpret. This is especially inconvenient when active substances are in short supply. Clearly, the growth factor-like substances extracted from *C. cinereus* (and from the fungi of previous studies) must be purified, identified and ideally, chemically synthesised so that their activity can be verified. Once the structure of DIF was determined (Morris *et al.*, 1987), its exact function during the differentiation process in *Dictyostelium discoideum* was verified (Williams *et al.*, 1987) and great strides were made in elucidating the binding proteins/receptors with which it interacted (Insall & Kay, 1990) to effect a developmental change in amoebae. Purified Fungiflex would provide a tool to establish which cells produce growth factors, their target(s), means of transport, mode of action, and cross-reactivity.

We are at the infancy in our knowledge of how mushrooms differentiate but the facilities and techniques for dissecting this question are at hand and we should adapt as many as are appropriate to the fungal system from the other embryological studies that are currently ongoing. The study of fungal growth factors and hormones promises to reveal a whole new world in mycology: understanding the interactions between neighbouring hyphae within a multihyphal structure, be it a sexual fruit body or a vegetative structure; understanding how the processes in the 'black box' between dikaryon formation and the development of a mature fruit body are coordinated; and opening up the possibilities in using these substances in commercial applications.

Acknowledgements

I thank the Natural Sciences and Engineering Council of Canada for a post-graduate scholarship, the British Mycological Society for funds to research Fungiflex and David Moore for helpful discussions on this topic.

References

Banuett, F. (1992). *Ustilago maydis*, the delightful blight. *Trends in Genetics* **8**, 174–180.

Berks, M., Traynor, D., Carrin, I., Insall, R. H. & Kay, R. R. (1991). Diffusible signal molecules controlling cell differentiation and patterning in *Dictyostelium*. *Development Supplement* **1**, 131–139.

Bölker, M. & Kahmann, R. (1993). Sexual pheromones and mating responses in fungi. *Plant Cell* **5**, 1461–1469.

Brickell, P. M. & Tickle, C. (1989). Morphogens in chick limb development. *BioEssays* **11**, 145–149.

Butler, G. M. (1995). Induction of precocious fruiting by a diffusible sex factor in the polypore *Phellinus contiguus*. *Mycological Research* **99**, 325–329.

Casselton, L. A. & Kües, U. (1994). Mating-type genes in homobasidiomycetes. In *The Mycota I. Growth, Differentiation and Sexuality* (ed. K. Esser & P. A. Lemke), pp. 307–321. Springer-Verlag: Berlin.

Chiu, S. W. & Moore, D. (1988). Evidence for developmental commitment in the differentiating fruit body of *Coprinus cinereus*. *Transactions of the British Mycological Society* **90**, 247–253.

Chiu, S. W. & Moore, D. (1990). Sporulation in *Coprinus cinereus*: use of an *in vitro* assay to establish the major landmarks in differentiation. *Mycological Research* **94**, 249–253.

Craig, G. D., Gull, K. & Wood, D. A. (1977). Stipe elongation in *Agaricus bisporus*. *Journal of General Microbiology* **102**, 337–347.

Duntze, W., Betz, R. & Nientiedt, M. (1994). Pheromones in yeasts. In *The Mycota I. Growth, Differentiation and Sexuality* (ed. K. Esser & P. A. Lemke), pp. 381–399. Springer-Verlag: Berlin.

Dyer, P. S., Ingram, D. S. & Johnstone, K. (1992). The control of sexual morphogenesis in the ascomycetes. *Biological Reviews* **67**, 421–458.

Eilers, F. I. (1974). Growth regulation in *Coprinus radiatus*. *Archives of Microbiology* **96**, 353–364.

Flegg, P. B. & Wood, D. A. (1985). Growth and fruiting. In *The Biology and Technology of the Cultivated Mushroom* (ed. P. B. Flegg, D. M. Spencer & D. A. Wood), pp. 141–177. John Wiley & Sons: New York.

Glass, N. L. & Nelson, M. A. (1994). Mating-type genes in mycelial ascomycetes. In *The Mycota I. Growth, Differentiation and Sexuality* (ed. K. Esser & P. A. Lemke), pp. 295–305. Springer-Verlag: Berlin.

Gooday, G. W. (1982). Metabolic control of fruitbody morphogenesis in *Coprinus cinereus*. In *Basidium and Basidiocarp: Evolution, Cytology,*

Function, and Development (ed. K. Wells & E. K. Wells), pp. 157–173. Springer-Verlag: Berlin.

Gooday, G. W. (1985). Elongation of the stipe of *Coprinus cinereus*. In *Developmental Biology of Higher Fungi, Vol. 10, British Mycological Society Symposium Series* (ed. D. Moore, L. A. Casselton, D. A. Wood & J. C. Frankland), pp. 311–331. Cambridge University Press: Cambridge.

Gooday, G. W. (1994). Hormones in mycelial fungi. In *The Mycota, I, Growth, Differentiation and Sexuality* (ed. J. G. H. Wessels & F. Meinhardt), pp. 401–411. Springer-Verlag: Berlin & Heidelberg.

Gooday, G. W. & Adams, D. J. (1993). Sex hormones and fungi. *Advances in Microbial Physiology* **34**, 69–145.

Greening, J. P., Holden, J. & Moore, D. (1993). Distribution of mechanical stress is not involved in regulating stipe gravitropism in *Coprinus cinereus*. *Mycological Research* **97**, 1001–1004.

Greening, J. P. & Moore, D. (1996). Morphometric analysis of cell size patterning involved in regulating stem gravitropism in *Coprinus cinereus*. *Advances in Space Research* **17**, 83–86.

Gruen, H. E. (1963). Endogenous growth regulation in carpophores of *Agaricus bisporus*. *Plant Physiology* **38**, 652–666.

Gruen, H. E. (1969). Growth and rotation of *Flammulina velutipes* fruitbodies and the dependence of stipe elongation on the cap. *Mycologia* **61**, 149–166.

Gruen, H. E. (1976). Promotion of stipe elongation in *Flammulina velutipes* by a diffusate from excised lamellae supplied with nutrients. *Canadian Journal of Botany* **54**, 1306–1315.

Gruen, H. E. (1982). Control of stipe elongation by the pileus and mycelium in fruitbodies of *Flammulina velutipes* and other Agaricales. In *Basidium and Basidiocarp: Evolution, Cytology, Function, and Development* (ed. K. Wells & E. K. Wells), pp. 125–155. Springer-Verlag: Berlin.

Guilfoyle, T. J., Hagen, G., Li, Y., Ulmasov, T., Liu, Z-B., Strabala, T. & Gee, M. (1993). Auxin-regulated transcription. *Australian Journal of Plant Physiology* **20**, 489–502.

Hagimoto, H. & Konishi, M. (1959). Studies on the growth of fruitbody of fungi. I. Existence of a hormone active to the growth of fruitbody in *Agaricus bisporus* (Lange) Sing. *Botanical Magazine of Tokyo* **72**, 359–366.

Hagimoto, H. & Konishi, M. (1960). Studies on the growth of fruit body of fungi. II. Activity and stability of the growth hormone in the fruit body of *Agaricus bisporus* (Lange) Sing. *Botanical Magazine of Tokyo* **73**, 283–287.

Hammad, F., Ji, J., Watling, R. & Moore, D. (1993). Cell population dynamics in *Coprinus cinereus*: co-ordination of cell inflation throughout the maturing basidiome. *Mycological Research* **97**, 269–274.

Hammad, F., Watling, R. & Moore, D. (1993). Cell population dynamics in *Coprinus cinereus*: narrow and inflated hyphae in the basidiome stipe. *Mycological Research* **97**, 275–282.

Insall, R. & Kay, R. R. (1990). A specific DIF binding protein in *Dictyostelium*. *EMBO Journal* **9**, 3323–3328.

Ip, Y. T., Levine, M. & Small, S. J. (1992). The *bicoid* and *dorsal* morphogens use a similar strategy to make stripes in the *Drosophila* embryo. *Journal of Cell Science, Supplement* **16**, 33–38.

Kamada, T. (1994). Stipe elongation in fruit bodies. In *The Mycota I. Growth, Differentiation and Sexuality* (ed. K. Esser & P. A. Lemke), pp. 367–379. Springer-Verlag: Berlin.

Kämper, J., Bölker, M. & Kahmann, R. (1994). Mating-type genes in heterobasidiomycetes. In *The Mycota I. Growth, Differentiation and Sexuality* (ed. K. Esser & P. A. Lemke), pp. 323–331. Springer-Verlag: Berlin.

Kawai, G. & Ikeda, Y. (1982). Fruiting-inducing activity of cerebrosides observed with *Schizophyllum commune*. *Biochimica et Biophysica Acta* **719**, 612–618.

Kay, R. R., Berks, M. & Traynor, D. (1989). Morphogen hunting in *Dictyostelium*. *Development* **107** Supplement, 81–90.

Kelly, M., Burke, J., Smith, M., Klar, A. & Beach, D. (1988). Four mating-type genes control sexual differentiation in the fission yeast. *EMBO Journal* **7**, 1537–1547.

Kemp, R. F. O. (1977). Oidial homing and the taxonomy and speciation of basidiomycetes with special reference to the genus *Coprinus*. In *The Species Concept in Hymenomycetes* (ed. H. Clemençon), pp. 259–273. Cramer: Vaduz.

Key, J. L. (1989). Modulation of gene expression by auxin. *BioEssays* **11**, 52–58.

Kher, K., Greening, J. P., Hatton, J. P., Novak Frazer, L. & Moore, D. (1992). Kinetics and mechanics of stem gravitropism in *Coprinus cinereus*. *Mycological Research* **96**, 817–824.

Konishi, M. (1967). Growth promoting effect of certain amino acids on the *Agaricus* fruit body. *Mushroom Science* **6**, 121–134.

Konishi, M. & Hagimoto, H. (1961). Studies on the growth of fruitbody of fungi. III. Occurrence, formation and destruction of indole acetic acid in the fruitbody of *Agaricus bisporus* (Lange) Sing. *Plant and Cell Physiology* **2**, 425–434.

Kozak, K. R. & Ross, I. K. (1991). Signal transduction in *Coprinus congregatus*: evidence for the involvement of G proteins in blue light photomorphogenesis. *Biochemical and Biophysical Research Communications* **179**, 1225–1231.

Kües, U. & Casselton, L. A. (1992). Fungal mating type genes – regulators of sexual development. *Mycological Research* **96**, 993–1006.

Kües, U. & Casselton, L. A. (1994). Mating-type genes in homobasidiomycetes. In *The Mycota I. Growth, Differentiation and Sexuality* (ed. K. Esser & P. A. Lemke), pp. 307–321. Springer-Verlag: Berlin.

Kües, U., Asante-Owusu, R. N., Mutasa, E. S., Tymon, A. M., Pardo, E. H., O'Shea, S. F., Göttgens, B. & Casselton, L. A. (1994). Two classes of homeodomain proteins specify the multiple *A* mating types of the mushroom *Coprinus cinereus*. *Plant Cell* **6**, 1467–1475.

Kües, U., Richardson, W. V. J., Tymon, A. M., Mutasa, E. S., Göttgens, B., Gaubatz, S., Gregoriades, A. & Casselton, L. A. (1992). The combination of dissimilar alleles of the *Aα* and *Aβ* gene complexes, whose proteins contain homeo domain motifs, determines sexual development in the mushroom *Coprinus cinereus*. *Genes & Development* **6**, 568–577.

Lawrence, P. A. (1992). *The Making of a Fly*. Blackwell Scientific: Oxford, U.K.

Löbler, M. & Klämbt, D. (1985). Auxin-binding protein from coleoptile membranes of corn (*Zea mays* L.). II. Localisation of a putative auxin receptor. *Journal of Biological Chemistry* **260**, 9854–9859.

Manachère, G. (1988). Regulation of sporophore differentiation in some macromycetes, particularly in Coprini: an overview of some experimental studies, from fruiting initiation to sporogenesis. *Cryptogamie Mycologie* **9**,

291–323.

Matthews, T. R. & Niederpruem, D. J. (1972). Differentiation in *Coprinus lagopus*. I. Control of fruiting and cytology of initial events. *Archives of Microbiology* **87**, 257–268.

Mau, J-L., Beelman, R. B. & Ziegler, G. R. (1992). Effect of 10-oxo-*trans*-8-decenoic acid on growth of *Agaricus bisporus*. *Phytochemistry* **31**, 4059–4064.

Michalenko, G. O. (1971). The assay of growth-promoting substances in *Flammulina velutipes* fruitbodies with a standardised stipe curvature test. Ph.D. Thesis, University of Saskatchewan, 213 pp.

Moore, D. (1984a). Developmental biology of the *Coprinus cinereus* carpophore: metabolic regulation in relation to cap morphogenesis. *Experimental Mycology* **8**, 283–297.

Moore, D. (1984b). Positional control of development in fungi. In *Positional Controls in Plant Development* (ed. P. W. Barlow & D. J. Carr), pp. 107–135. Cambridge University Press: Cambridge.

Moore, D. (1991). Perception and response to gravity in higher fungi – a critical appraisal. *New Phytologist* **117**, 3–23.

Moore, D. (1995). Tissue formation. In *The Growing Fungus* (ed. N. A. R. Gow & G. M. Gadd), pp. 423–465. Chapman & Hall: London.

Moore, D., Elhiti, M. M. Y. & Butler, R. D. (1979). Morphogenesis of the carpophore of *Coprinus cinereus*. *New Phytologist* **83**, 695–722.

Moore, D., Greening, J. P., Hatton, J. P. & Novak Frazer, L. (1994). Gravitational biology of mushrooms: a flow-chart approach to characterising processes and mechanisms. *Microgravity Quarterly* **4**, 21–24.

Moore, D., Hock, B., Greening, J. P., Kern, V. D., Novak Frazer, L. & Monzer, J. (1996). Gravimorphogenesis in agarics. *Mycological Research* **100**, 257–273.

Moore, D. & Novak Frazer, L., Victoria University of Manchester. (1993). Controlling fungal growth. U. K. Patent Application Number 9303073.2.

Moore, D. & Novak Frazer, L., Victoria University of Manchester. (1994). Controlling fungal growth. U. K. Patent Application Number 9403007.9.

Morris, H. R., Taylor, G. W., Masento, M. S., Jermyn, K. A. & Kay, R. R. (1987). Chemical structure of the morphogen differentiation inducing factor from *Dictyostelium discoideum*. *Nature* **328**, 811–814.

Mulder, G. H. & Wessels, J. G. H. (1986). Molecular cloning of RNAs differentially expressed in monokaryons and dikaryons of *Schizophyllum commune* in relation to fruiting. *Experimental Mycology* **10**, 214–227.

Mullins, J. T. (1994). Hormonal control of sexual dimorphism. In *The Mycota I. Growth, Differentiation and Sexuality* (ed. K. Esser & P. A. Lemke), pp. 413–421. Springer-Verlag: Berlin.

Nicola, N. A. (1994). *Guidebook to the Cytokines and their Receptors*. Oxford University Press: Oxford, U.K.

Pegg, G. F. (1973). Gibberellin-like substances in the sporophores of *Agaricus bisporus* (Lange) Imbach. *Journal of Experimental Botany* **24**, 675–688.

Reijnders, A. F. M. & Moore, D. (1985). Developmental biology of agarics- an overview. In *Developmental Biology of Higher Fungi, Vol. 10, British Mycological Society Symposium Series* (ed. D. Moore, L. A. Casselton, D. A. Wood & J. C. Frankland), pp. 581–595. Cambridge University Press: Cambridge.

Robert, J. C. (1990). On the light control of fruit-body development in *Coprinus congregatus*. In *Fourth International Mycological Congress Abstracts* (ed. A. Reissinger & A. Bresinsky), pp. 91. University of Regensberg: Regensberg.

Robert, J. C. & Bret, J. P. (1987). Release of an inhibitor of stipe elongation from illuminated caps of *Coprinus congregatus* mushrooms. *Canadian Journal of Botany* **65**, 505–508.

Rosin, I. V., Horner, J. & Moore, D. (1985). Differentiation and pattern formation in the fruit body cap of *Coprinus cinereus*. In *Developmental Biology of Higher Fungi, Vol. 10, British Mycological Society Symposium Series* (ed. D. Moore, L. A. Casselton, D. A. Wood & J. C. Frankland), pp. 333–351. Cambridge University Press: Cambridge.

Ross, I. K. (1985). Determination of the initial steps in differentiation in *Coprinus congregatus*. In *Developmental Biology of Higher Fungi Vol. 10, British Mycological Society Symposium Series* (ed. D. Moore, L. A. Casselton, D. A. Wood & J. C. Frankland), pp. 353–373. Cambridge University Press: Cambridge, U.K.

Rusmin, S. & Leonard, T. J. (1978). Biochemical induction of fruiting in *Schizophyllum*. Isolation and preliminary purification of an inducing substance from *Agaricus bisporus* mushrooms. *Plant Physiology* **61**, 538–543.

Sachs, T. (1991). Cell polarity and tissue patterning in plants. *Development Supplement* **1**, 83–93.

Salisbury, F. B. & Ross, C. W. (1985). *Plant Physiology*. Wadsworth Publishing Company: Belmont.

Staben, C. (1995). Sexual reproduction in higher fungi. In *The Growing Fungus* (ed. N. A. R. Gow & G. M. Gadd), pp. 383–422. Chapman & Hall: London.

St Johnston, D. & Nüsslein-Volhard, C. (1992). The origin of pattern and polarity in the *Drosophila* embryo. *Cell* **68**, 201–219.

Tickle, C. (1991). Retinoic acid and chick limb development. *Development Supplement* **1**, 113–121.

Tressl, R., Bahri, D & Engel, K-H. (1982). Formation of eight-carbon and ten-carbon components in mushrooms (*Agaricus campestris*). *Journal of Agricultural and Food Chemistry* **30**, 89–93.

Tymon, A. M., Kües, U., Richardson, W. V. J. & Casselton, L. A. (1992). A fungal mating type protein that regulates sexual and asexual development contains a POU-related domain. *EMBO Journal* **11**, 1805–1813.

Ulmasov, T., Liu, Z-B., Hagen, G. & Guilfoyle, T. J. (1995). Composite structure of auxin responsive elements. *Plant Cell* **7**, 1611–1623.

Uno, I. & Ishikawa, T. (1971). Chemical and genetic control of induction of monokaryotic fruiting bodies in *Coprinus macrorhizus*. *Molecular and General Genetics* **113**, 228–239.

Uno, I. & Ishikawa, T. (1973a). Purification and identification of the fruiting inducing substances in *Coprinus macrorhizus*. *Journal of Bacteriology* **113**, 1240–1248.

Uno, I. & Ishikawa, T. (1973b). Metabolism of adenosine 3',5'-cyclic monophosphate and induction of fruiting bodies in *Coprinus macrorhizus*. *Journal of Bacteriology* **113**, 1249–1255.

Uno, I. & Ishikawa, T. (1982). Biochemical and genetic studies on the initial events of fruitbody formation. In *Basidium and Basidiocarp: Evolution, Cytology, Function, and Development* (ed. K. Wells & E. K. Wells), pp. 113–123. Springer-Verlag: Berlin.

Urayama, T. (1969). Stimulative effect of extracts from fruit bodies of *Agaricus bisporus* and some other hymenomycetes on primordium formation in *Marasmius* sp. *Transactions of the Japanese Mycological Society* **10**, 73–78.

van der Valk, P. & Marchant, R. (1978). Hyphal ultrastructure in fruit-body primordia of the basidiomycetes *Schizophyllum commune* and *Coprinus cinereus*. *Protoplasma* **95**, 57–72.

Watkinson, S. C. (1979). Growth of rhizomorphs, mycelial strands, coremia and sclerotia. In *Fungal Walls and Hyphal Growth, Vol.* **2**, *British Mycological Society Symposium Series* (ed. J. H. Burnett & A. P. J. Trinci), pp. 93–113. Cambridge University Press: Cambridge.

Watling, R. & Moore, D. (1994). Moulding moulds into mushrooms: shape and form in the higher fungi. In *Shape and Form in Plants and Fungi, No.* **16**, *Linnean Society Symposium Series* (ed. D. S. Ingram & A. Hudson), pp. 271–290. Academic Press: London.

Wessels, J. G. H. (1992). Gene expression during fruiting in *Schizophyllum commune*. *Mycological Research* **96**, 609–620.

Wessels, J. G. H. (1993a). Fruiting in the higher fungi. In *Advances in Microbial Physiology, Vol.* **34** (ed. A. H. Rose), pp. 147–202. Academic Press: San Diego.

Wessels, J. G. H. (1993b). Tansley review no. 45. Wall growth, protein excretion and morphogenesis in fungi. *New Phytologist* **123**, 397–413.

Wessels, J. G. H. (1994). Development in fruit bodies in homobasidiomycetes. In *The Mycota I. Growth, Differentiation and Sexuality* (ed. K. Esser & P. A. Lemke), pp. 351–365. Springer-Verlag: Berlin.

Willetts, H. J. & Bullock, S. 1992. Developmental biology of sclerotia. *Mycological Research* **96**, 801–816.

Williams, J. G., Ceccarelli, A., McRobbie, S., Mahbubani, H, Kay, R. R., Early, A., Berks, M. & Jermyn, K. A. (1987). Direct induction of *Dictyostelium* prestalk gene expression by DIF provides evidence that DIF is a morphogen. *Cell* **49**, 185–192.

Yashar, B. M. & Pukkila, P. J. (1985). Changes in polyadenylated RNA sequences associated with fruiting body morphogenesis in *Coprinus cinereus*. *Transactions of the British Mycological Society* **84**, 215–226.

Chapter 8

Patterns in fungal development – fruiting patterns in nature

ROY WATLING

Summary

Three modifications of the basic umbrella-shape (namely a cap surmounting a stem and protecting the spore-producing layers) which are found in the agaric fungi are discussed. They probably appeared several times in the evolution of the agaricoid fungi, probably in response to ecological factors and biological strategies. Parallel pileate anamorphic stages of agarics are also discussed. The suitability of the cyphelloid and pleurotoid form for developmental studies is emphasised.

Introduction

The agarics and boletes range from the tiny fruit bodies (basidiomes) generally associated with species of *Marasmiellus*, *Marasmius* and *Mycena* and their allies (Tricholomataceae; Fig. 1), which are not more than 3-6 mm in diameter, to the enormous boletes in the genus *Phlebopus* (Gyrodontoideae: Boletaceae; Fig. 2) which may measure over 350 mm in diameter and weigh 20 kg. The heavy and compact structures of such boletes have been considered primitive (Corner, 1993) contrasting with the delicate fruit bodies of *Leucocoprinus fragillissimus* (Fig. 3) and the deliquescent fruit bodies of agarics placed in the genus *Coprinus*. Whatever their size there is a general similarity in form; that is, the umbrella-shape composed of a pileus or cap, surmounting a stipe or stem (Figs 4–6 and Fig. 43a), and this has been considered a most fundamental structure by Corner, being derived from a club or coralloid form (Figs 7 & 8; Corner, 1964; 1972); judging from recent finds, this structure is also rather ancient (Hibbett, Grimaldi & Donoghue, 1995). The pores or gills (lamellae) in which or on which the basidiospores develop on spore-bearing tissue (hymenophore) form be-

Fig. 1. *Marasmius rotula*, a woodland saprotroph on fallen twiggy debris 20–50 mm in height.

neath the protective cap; the former is the bolete with its spongy lower surface and the latter the cultivated mushroom, each related to one another by their putrescent, rather than perennial fruit body. However, when examined closely a few modifications to this basic pattern can be found. The overwhelming dominance of the classic mushroom shape has sculptured the early classification of the mushrooms and toadstools so much that mycologists have been blinded until recently to the natural relationships with other basidiomycetous fungi. In many ways, what is happening in mycology now is reminiscent of the impact of comparative morphology on understanding relationships between animals which occurred in the nineteenth century.

Pleurotoid agarics

The commonest modification recognized in the classic literature was the fruit body which lacked or had a reduced, eccentric stem; such structures were termed pleurotoid fruit bodies (Figs 9–12). They are clearly seen in such genera as *Crepidotus* (brown-spores), *Melanotus* (purple-black spores),

Fig. 2. *Phlebopus sudanicus*, a dry sclerophyll ectotroph which can exceed 125 mm in cap diameter.

Clitopilus & *Claudopus* (pink spores), and *Pleurotus* (white or cream-coloured or pale liliaceous spores, Fig. 13), genera now considered to be referrable to quite different families. The loss of the stem has therefore occurred in different groups and most probably at several different times in evolution. The majority of such agarics would have been placed in a single family in older texts. Their growth pattern may be a response to growth on woody substrates, herbaceous stems and the like, or vertical surfaces such as trail verges and woodland banks (Tables 1 & 2), though not all fungi found on vertical surfaces follow this strategy. Some possess a 'normal', central stem which may commence growth horizontally but soon turns upwards in response to gravity in the classic negatively gravitropic response (Moore *et*

Fig. 3. *Coprinus plicatilis* a terricolous grassland saprotroph 25–75 mm in height.

al., 1996). Several fungi, again in all the major families, use this strategy or exhibit an eccentric or slightly off-centre stem. Stem eccentricity might be simply a response to a local, usually physical, condition which requires adjustment of the growth form. Thus an agaric will respond by distorting its basic shape unlike a species of *Hydnellum* or *Heterobasidion* (Aphyllophorales) which might even incorporate the obstacle in order to retain the vertical alignment (Fig. 14). In a few, considered more primitive on other criteria, such as *Gyrodon rompellii* and *G. exiguus*, eccentric stems are present from the very beginning. This contrasts with *Pleurotus ostreatus* which in culture (Figs 9a–h) commences centrally stipitate. An extension of this same phenomenon can be found in the closely-related taxa like

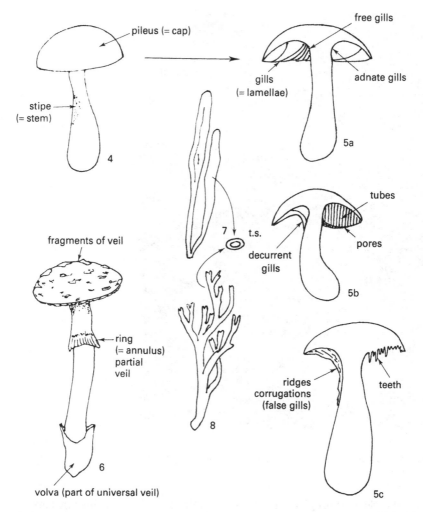

Figs 4–8. Fruit body morphology and structure. **Fig. 4.** Mushroom fruit body. **Fig. 5a.** Section of mushroom fruit body showing gill attachment; free if not attached to stem apex and adnate if attached to the stem for over half their length. **Fig. 5b.** As 5a but showing gills running down the stem (decurrent) or replaced by tubes as in the boletes. **Fig. 5c.** As 5a but gills replaced by wrinkles as in *Cantharellus* or by 'teeth' as in hydnaceous fungi. **Fig. 6.** Mushroom fruit body (*Amanita muscaria*) with fragments of volva on cap and remnants at base of stem, and with ring formed from remains of partial veil. **Fig. 7.** Fairy Club fruit body with amphigenous hymenium. **Fig. 8.** Coralloid fruit body with amphigenous hymenium.

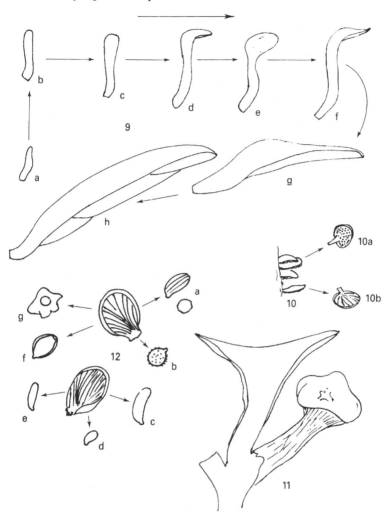

Figs 9–12. Fruit body morphology and structure. **Fig. 9a–h.** Development of eccentric fruit body of *Pleurotus ostreatus* (from culture). **Fig. 10.** *Panellus stypticus.* **Fig. 10a & b.** Undersurface of *Dictyopanus pusillus* (pores) and *Panellus mitis* (gills). **Fig. 11.** Decurrent gills of *Pleurotus cornucopiae* on central stipitate fruit body. **Fig. 12.** Pleurotoid fruit body – sporograms: clockwise *Clitopilus* (a), *Crepidotus* (b), *Pleurotus* (c), *Panus* (d) *Pleurotellus* (e), *Melanotus* (f), *Claudopus* (g).

Fig. 13. *Pleurotus ostreatus*: lignicolous saprotroph with eccentric stem on cut stump; sometimes a weak parasite.

P. cornucopiae which retains the central stem even into maturity (Fig. 11). Careful examination of the majority of *Melanotus* spp. and *Crepidotus* spp. reveals a stem at an early stage which, because of little or no elongation, is then squeezed out by expansion of the cap. In some species the stem remains, even in mature material, as a curved, eccentric attachment.

In Europe, on a single fungus foray few of the fungi recorded have eccentric stems. However, in their particular ecological niche several score of individuals may be found. In the Tricholomataceae in Britain, for instance, 4.5% of the constituent taxa lack stems (Tables 1 & 2). The mushroom shape, therefore, is dominant in a whole series of families, not all of which are closely related to each other.

Certainly abnormalities can be found in the field but are usually explained as being due to attack by a parasite or other effects leading to confusion in the orientation and development of tissues. This occurs particularly in the gills of certain Russulaceae which, when attacked by members of the genus *Hypomyces* (Hypomycetaceae; Ascomycotina), react by reduced elongation of the gills and their replacement or permeation by parasite mycelium. Similarly swelling of the stem and miniaturization of the cap occurs in some boletes e.g. *Boletus edulis*, when attacked by *Sepedonium chrysospermum*. The gills in Russulaceae attacked by *Hypomyces* support the developing perithecia of the ascomycete whereas in *Boletus* in Britain the fruit body is converted to a mass of yellow asexual chlamydospores. In other parts of the temperate world the teleomorph (sexual stage) of the same fungus is produced and then replaces the poroid hymenophore of the bolete.

Table 1. *Agarics with lateral stem assigned to their natural groupings: many were previously placed in the single artificial family Pleurotaceae*

Agaric groupings			Pleurotoid element	Cyphelloid
Entolomatales	Entolomataceae	*Leptonia*	*Claudopus*	
		Eccilia		
		Clitopilus	*Clitopilus hobsonii*	
Cortinariales	Cortinariaceae	*Pholiota*	*Pleuroflammula*	
		Gymnopilus	*Pyrrhoglossum*	
	Crepidotaceae	*Tubaria*[1]	*Crepidotus*	*Episphaeria*
			Pleurotellus	*Chromocyphella*
		Simocybe[1]	*S. amazonica* & *S. fulvofibrillosa*	
	Strophariaceae	*Psilocybe*	*Melanotus*	
	Russulaceae	*Lactarius* (Sect. *Panuoidei*)	*L. panuoides* = *Pleurogala*	
	Paxillaceae	*Paxillus*	*Tapinella panuoides Omphalotus*	
	Boletaceae *sensu lato* (including Gyrodontaceae)	*Pulveroboletus Gyrodon* (*Boletinellus* spp.)	*P. acaulis* *G. exiguus*	
	Pleurotaceae[2]	*Lentinus/Panus*	*Pleurotus Phyllotopsis*	

[1] Placed here by Singer
[2] Considered by Singer to be part of the Polyporaceae

Stomach fungi & hypogeous elements

The 'field mycologist' generally recognises a puffball or one of its close relatives (Fig. 15) because they possess a characteristic enclosed (endocarpic) development resulting in their spores being borne inside the fruit body. Many agarics pass through this stage during their maturation (Reijnders, 1948, 1963) with the spore bearing tissue hidden from the environment in the earliest stages. Fayod (1889) recognised this and termed such fungi which started enclosed and later became exposed to the environment hemiangiocarpic (Figs 16–18) in contrast to both the endocarpic (= angiocarpic; Figs 20–22) gasteromycetes and the gymnocarpic agarics where all the stages are unprotected (Reijnders, 1963; Watling, 1985a; Figs 19a–e). Even in the latter case further protection of the rather delicate

Table 2. *Tricholomataceous agarics with lateral stem assigned to their natural groupings: many were previously placed in the single artificial family Pleurotaceae*

Agaric groupings			Pleurotoid element	Cyphelloid
Favolaschiaceae			*Favolaschia*	
Tricholomataceae	Tricholomateae	Subtribe Omphalinae *Omphalina & Gerronema*	*Pleurocollybia*	*Fissolimbus Leptoglossum* (= *Arrhenia*)
	Collybieae	*Collybia*	*Pleurocybella Cheimonophyllum Anthracophyllum*	
	Leucopaxilleae	*Clitocybula C. lacerata*	*C. tarnensis*	
	Lyophylleae	*Lyophyllum*	*Hypsizygus*	
	Marasmieae	Subtribe Marasminae *Marasmius* Sect. Neosessiles	*M. polycystis*	
		Subtribe Crinipellinae *Crinipellis*	*Amyloflagellula Chaetocalathus*	
	Resupinateae	*Agaricochaete*	*Resupinatus*	*Hohenbuehelia*
	Myceneae	*Mycena*		*Delicatula Hemimycena* (= *Helotium*)
		Baeospora B. myosura	*B. pleurotoides Dictyopanus Panellus*	
	Panelleae			
	Rhodoteae		*Rhodotus*	

hymenial surfaces is provided by the production of secondary protective membranes (veils) in some species either by a woolly growth of the cap margin, e.g. *Lactarius torminosus* or by membranes which hug the stem as in some species of *Suillus* (Watling, 1985b).

The gasteromycetes include fungi with a spectrum of fruit body shapes and structures, all being brought together in a single group because of their endocarpic development. It is now considered quite an artificial grouping although classically it has been accepted and split into the botanically required genera, sections of genera, families and orders etc. Such a systematic arrangement is still a convenient handle by which one can group such fungi but arrayed amongst them are taxa which in section even in adulthood look extremely similar to agarics. Such 'Peter Pan' fungi have, perhaps by historical accident, been brought together in the single family Secotiaceae, which by microscopic analysis can be shown quickly and

Fig. 14. *Hydnellum caeruleum*: conifer woodland ectotroph engulfing surrounding vegetation.

convincingly to be an artificial assemblage (Figs 26–30). Thus species with basidiospores which become blue-black in solutions containing iodine (amyloid) and having flesh composed of filaments and pockets of rounded cells (Russulaceae; Figs 26 & 27) or ornamented, orange tawny pigmented basidiospores lacking a germ-pore and regularly arranged tissues in the gills (Figs 25 & 28; Cortinariaceae) or elongate ellipsoid shaped basidiospores and tubular hymenophore (Boletaceae; Fig. 30), etc., are all found. The secotioid fungi have proved very provocative and their study has spawned many publications, often with totally different interpretations, e.g. Singer & Smith (1960) and Thiers (1984). The secotioid fungi is such an artificial assemblage that taxa previously placed in the same family have probably evolved from adjacent subgenera or sections of genera and are less related to each other than they are to agaricoid members even in those same subgenera or sections (Table 3; Watling, 1978; Bougher, Tommerup & Malajcuk, 1993).

Fig. 15. Undescribed hypogeous basidiomycete (Hymenogastrales) from lowland dipterocarp forest of Malaysia.

The secotioid habit has undoubtedly evolved in a number of different environments to overcome rather different problems ranging from arid conditions (*Galeropsis*), snow-bank conditions (*Nivatogastrium*), or spores dispersed by molluscs in the absence of a small mammal fauna (*Clavatogaster* and *Weraroa* in New Zealand). The trophic strategy is generally the same as the closest agaric. For example, *Nivatogastrium* on wood like *Pholiota*, *Thaxterogaster* and *Macowanites* are ectomycorrhizal partners like their allies in Cortinariaceae and Russulaceae, respectively. The sizes and external characters of the fruit body parallel their apparently most closely related agaric, and similarities in secondary metabolites can be detected also (Table 3). A start has been made to study these supposed connections at the molecular level in the coprinoid fungi (Hopple & Vilgalys, 1994) and Boletales (Baura, Szaro & Bruns, 1992).

Alas, it is only the secondary metabolites and microscopic elements which can sort out the hypogeous gasteromycetes because here the basidiospores are borne endocarpically in a much reduced, potato-like fruit body with little if any differentiation into rudimentary cap, stem and gills (Figs 23–25)! Many look alike until dissected, whereupon their true relationships are revealed, although a few do show some similarities to their suggested relatives. Again, most of the relationships are with agarics and

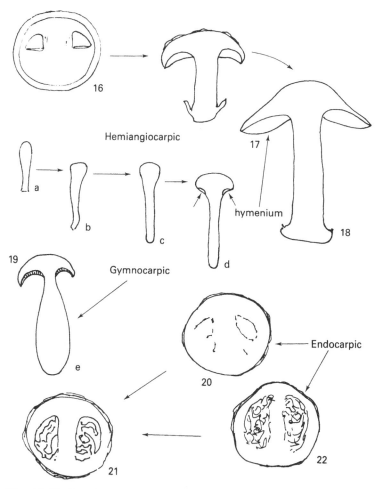

Figs 16–22. Developmental strategies. **Figs 16–18.** Hemiangiocarpic development in *Cortinarius* sp. Sections of fruit body showing hymenium exposed at maturity. **Fig. 19 a–e.** Gymnocarpic development in *Boletus* sp. with hymenium exposed at all times. **Figs 20–22.** Endocarpic development in secotioid fruit body showing totally enclosed hymenium.

range across a spectrum of families. Here the change has been almost certainly to exploit animal dispersal agents, thereby rejecting reliance on the wind. In both the secotioid and hypogeous fungi the comparative numbers of taxa are few (Singer, 1958), but contrary to what is often suggested they are certainly not absent in the tropics; the author has

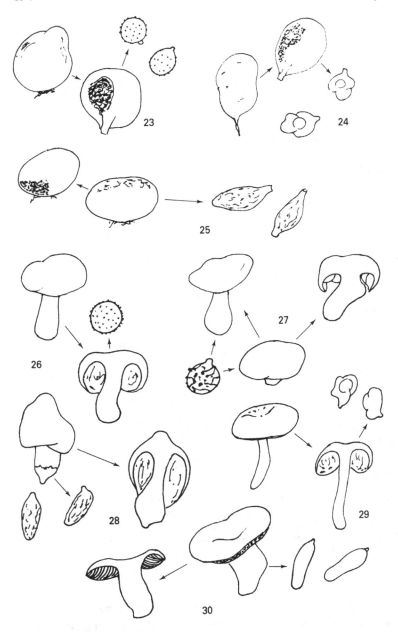

Figs 23–25. Hypogeous fungi. **Fig. 23.** *Hydnangium* – (*Laccaria*) Tricholomataceae.
Fig. 24. *Rhodogaster* – Entolomataceae. **Fig. 25.** *Hymenogaster* – Cortinariaceae.
Figs 26–30. Secotioid fungi. **Fig. 26.** *Macowanites* – Russulaceae. **Fig. 27.** *Elasmomyces* – Russulaceae. **Fig. 28.** *Thaxterogaster* – Cortinariaceae. **Fig. 29.** *Richonella* – Entolomataceae. **Fig. 30.** *Gastroboletus* – Boletaceae.

Table 3. *The Boletales and Russulales illustrate the broad range of hymenophore configurations which members of an order can span; they also have gasteroid relatives – a feature found throughout the agarics*

| Traditional taxa | Hymenogastrales | | Additional hymenophore configurations |
	Secotiaceae	Hymenogastraceae and Rhizopogonaceae	
Boletales, predominantly form poroid hymenophore			
Hygrophoropsidaceae			Agaricoid
Gyrodontaceae			Agaricoid (?)
Meiorganaceae			Agaricoid
Gomphidiaceae *Chroogomphus*			
Gomphidius	*Brauniellula*		
Paxillaceae *Paxillus*	*Gomphogaster*		
Boletaceae *Boletus*	*Austrogaster*		
Suillus	*Gastroboletus*		Agaricoid
Leccinum			
Phylloporus			
	Paxillogaster		
	Gymnopaxillus		
Chamonixiaceae *Gyroporus?*		*Chamonixia*	
Boletellaceae			Agaricoid
Strobilomycetaceae			
Coniophoraceae			Cantharelloid, hydnoid, merulioid, resupinate
Rhizopogonaceae		*Rhizopogon*	
Corneromycetaceae			Cantharelloid, hydnoid, merulioid, resupinate
Russulales, called the Astrogastraceous series, predominantly with gilled hymenophore			
Russulaceae *Lactarius*	*Arcangeliella*		Agaricoid and poroid
Russula	*Macowanites*		Agaricoid
	Elasmomyces	*Zelleromyces*	
Elasmomycetaceae		*Gymnomyces*	
		Martellia	

Taxa according to Julich (1981)

Table 4. *Fungi with cup-shaped fruit body arranged in their modern (natural) genera with affinities: Tricholomataceae is represented by tribes*

Cyphella spp.	Modern placement
C. alboviolascens	*Lachnella*: Marasmieae
C. anomala	*Cyphellopsis*: Cyphellopsidaceae
C. capula	*Calyptella*: Collybieae
C. digitalis	*Cyphella* : Cyphellaceae
C. filicina	*Nochascypha*: Marasmieae
C. fraxinicola	*Episphaeria*: Crepidotaceae
C. goldbachii (*C. lactea*)	*Cellypha*: Myceneae
C. mairei	*Cephaloscypha*: Marasmieae
C. minutissima	*Flagelloscypha*: Marasmieae
C. muscicola	*Chromocyphella*: Crepidotaceae
C. muscicola var. *neckerae*	*Mniopetalum* (*Rimbachia*) ; Collybieae
C. pallida	*Pellidiscus*: Crepidotaceae

collected *Corditubera* in West Africa and *Chamonixia* in Peninsula Malaysia.

Cyphelloid fungi

Gradually, other modifications are now being accepted within the circumscription of the agarics and boletes, but often still referred to as 'relatives' as if authors would prefer to forget them! Such a group is the cyphelloid fungi with cup-shaped fruit bodies (Figs 31–36). Whilst the umbrella-shape has been conserved in the majority of the agarics the cup-shaped fruit body has been strongly conserved in the ascomycetes, e.g. 52.5% taxa in the British mycota alone (Cannon *et al.*, 1985). Basidiomycetes with cup-shaped fruit bodies were all brought together in the classical literature in the genus *Cyphella* (Table 4; see Donk, 1966). Now they are to be found distributed in several families, e.g. Tricholomataceae, Strophariaceae (Table 5), indeed related to sections and subgenera in parallel to the secotioid fungi. Agerer (1983) has been instrumental in demonstrating a range of these relationships. Because of the lack of gills, all the cyphelloid fungi, irrespective of their agaric connections were placed in the Thelephoraceae in classic texts, the criterion being their smooth or wrinkled hymenium. In these texts, they joined *Stereum* and basidiomycetes which possessed a flattened, smooth or wrinkled plate-like fruit body called resupinate fungi, although as early as the turn of the century, Patouillard (1900) had indicated a new way forward. He demonstrated that some were related to the agarics whilst others showed affinities to the bracket fungi, or polypores and their allies. These

Table 5. *Examples of cyphelloid fungi assigned to their natural groupings of agaricoid morphotypes*

Agaricoid			Reduced or lacking gills, stipitate	Cyphelloid
Tricholomatales				
Tricholomataceae	Subtribe	*Omphalina*	*Leptoglossum*	*Cyphellostereum*
Tricholomateae	Omphalinae		*Arrhenia*	
Collybieae		*Collybia*	*Campanella*	*Calyptella*
		Trogia	*(Laschia)*	*Hispidocalyptella*
		Marasmiellus		*Incrustatocalyptella*
Resupinateae		*Agaricochaete*		*Stigmatolemma*
				Stromatoscypha
Marasmieae	Subtribe	*Marasmius*	*Gloiocephala*	
	Marasminae			
		Sect. *Alliacei*		
		M. alliaceus	*Hymenogloea*	
		Sect. *Epiphylli*	*M. pusillissimus*	
		M. epiphylloides		
		Sect. *Hygrometrici*		
		M. hudsonii	*Manuripia*	
		Sect. *Neosessiles*		
		M. sessilis		*Deigloria*
	Subtribe	*Crinipellis*		*Flagelloscypha*
	Crinipellinae	*Amyloflagellula*		*Lachnella*
Myceneae		*Mycena*	*Delicatula*	*Cellypha*
		Xeromphalina	*Hemimycena*	
		Fayodia		
Cortinariales				
Crepidotaceae		*Crepidotus*		
		Sect. *Crepidotus*		*Episphaeria*
		Sect.		*Chromocyphella*
		Echinosporae		

now find their place next to members of the Polyporales (Aphyllophorales; Table 6).

The majority of the fungi with cup-shaped fruit bodies related to the agarics are assignable to the Tricholomataceae. This is rather striking because in contrast to other agaric families this family lacks secotioid elements except for a very few exceptions (e.g. *Amparoina*). In Britain, cyphelloid fungi make-up 5.4% of the Tricholomataceae based on the British Checklist (Dennis, Orton & Hora, 1960), with three other taxa placed in the Crepidotaceae. Additionally, as indicated above, there are other taxa which are more related to the polypores (e.g. Schizophyllaceae) although considered members of the agarics in classical texts. Cyphelloid fungi comprise 4.5% of the world agaric mycota (Singer, 1986) but judging

Table 6. *Examples of cyphelloid taxa with natural affinities with*
polypore-like fungi

Corticioid – resupinate		Cyphelloid, cup-shaped	False 'agarics'
Aleurodiscales			
Aleurodiscaceae	*Dendrothele*	←*Aleurodiscus*	
Cyphellaceae	*Cerocorticium*	*Cyphella digitalis*	
Cytidiaceae	*Cytidia*	→with age	
Meruliales			
Auriculariopsidaceae		*Auriculariopsis*	
Meruliaceae			
Plicaturaceae		*Plicatura*	*Plicaturopsis*
Tricholomatales sensu *Julich, non Singer*			
Cyphellopsidaceae		*Calathella*	
		Cyphellopsis	See Collybieae,
		Merismodes	Table 5
Favolaschiaceae		← *Favolaschia* →	
(Related through			
Aleurodiscaceae)			
Stromatoscyphellaceae		*Stromatoscyphella*	

from observations in the tropics and more recent publications (e.g. Horak
& Desjarden, 1994) this proportion is sure to increase.

The cup-shaped fruit body is probably strongly tied to habitat exploitation; growing on sides of herbaceous stems, underside of twigs, on the basal
parts of plants, in much the same way as several bird's nest fungi, for
example, and most are probably saprotrophs. There are some exceptions;
e.g. *Calyptella campanula* which is parasitic in hydroponic cultures of
tomatoes (Fig. 41; Clark, Richardson & Watling, 1983) and *Favolaschia*
dybowskyana which is an orchid symbiont (Jonsson & Nyland, 1979).
Cyphelloid fungi are never very large and can be confused easily with
members of the Ascomycotina (Figs 37–40), especially Helotiales
(Leotiaceae and Hyaloscyphaceae), by field workers. In some of the
Pezizales the ascomata can become quite large and disk-shaped, e.g. *Peziza*
vesiculosa (Fig. 37), whilst in *Helvella* there is a morphological series from

Figs 31–40 (*opposite*). Cyphelloid fungi and agaricoid non-basidiomycetes. Fruit
bodies with sections. Fig. 31a & b. *Flagelloscypha* spp. fruit bodies and sections. Fig.
32. Section showing hymenium with basidia. Fig. 33. *Henningsomyces pubera* with
spores and branched diverticulate hairs from surface. Fig 34a & b. *Cyphellopsis*
anomala with capitate, septate hairs from surface and spores (34a);. tight cluster of
fruit bodies and spores (34b). Fig. 35. *Chromocyphella muscicola* with spore. Fig. 36.
Flagelloscypha citrispora, fruit body hairs and spores. Fig. 37. *Peziza vesiculosa*

(Ascomycotina: Pezizales). **Fig. 38**. *Baeomyces rufus* (Ascomycotina: Leotiales: lichenized discomycete). **Fig. 39**. *Cudoniella acicularis* (Ascomycotina: Leotiales: non-lichenized discomycete). **Fig. 40**. Section of hymenium showing asci.

Fig. 41. *Cyphella campanula* on living stems of tomato plants in hydroponic culture.

stipitate saddle-shaped to almost umbrella-shaped ascomata (Dissing, 1966). In the discomycetes, however, the agaric-like shape is hardly ever attained; *Cudoniella acicularis* (Fig. 39) is an example, as is the lichenized genus *Baeomyces* (Fig. 38).

Revision of the criteria used in separating larger fungi one from the other and showing what are considered natural relationships has undoubtedly been paramount in unravelling the various elements found within the cyphelloid fungi. Such revision has also had far reaching effects in the classification of the resupinate fungi, and those fungi which resemble 'fairy' clubs or corals, i.e. the clavarioid fungi (Table 7).

Fig. 42. *Cantharellus cibarius*, a widespread ectomycorrhizal fungus associated with a wide host-range.

Chanterelles and their relatives

For generations the chanterelles (*Cantharellus* spp.; Figs 42 & 43b) and their relatives have been classified with the agarics undoubtedly because of their similarity in basic form, being mushroom-shaped, with a cap, stem and gill-like folds. It was appreciated even in the classical literature that they differed from the common mushroom in that they possessed gill-folds and wrinkles and not true gill-plates but this was not thought significant. The tendency to regard chanterelles as agarics persists in some quarters even today but cytological and anatomical characters point to a closer relationship to the club- or coral-fungi and the large group of gill-less fungi including the brackets and some of those with spore-producing tissue on tooth-like structures.

The chanterelles all possess a gymnocarpic development, that is without the production of velar tissues, but careful examination particularly of the spore morphology indicates that even this group is artificial. Thus a cluster of taxa can be linked to the coralloid *Ramaria*, namely *Gomphus*, characterised by elongate, ornamented, pigmented spores (Figs 44–47) and a further

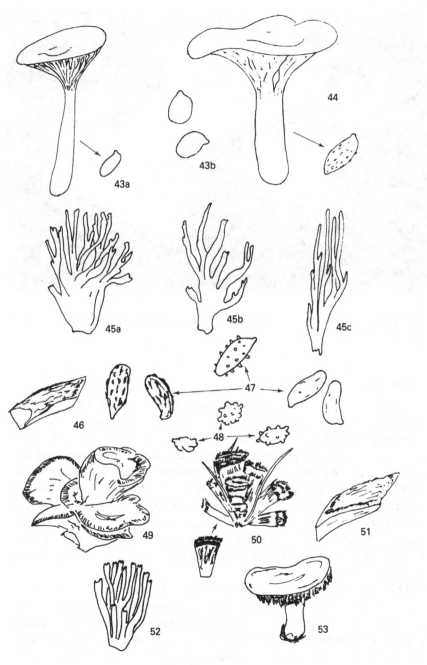

Table 7. *Fungi with club- and coral-shaped fruit body arranged in their modern (natural) genera with affinities as outlined by Knudsen (1995) for the Nordic Macromycetes*

Genus	Family	Order
Clavaria	Clavariaceae	Cantharellales
Clavulinopsis		
Ramariopsis		
Clavulina	Clavulinaceae	Cantharellales
Sparassis	Sparassidaceae	Cantharellales
Typhula	Typhulaceae	Cantharellales
Lachnocladium	Lachnocladiiaceae	Lachnocladiales
Clavicorona	Clavicoronaceae	Hericiales
Clavariadelphus	Clavariadelphaceae	Gomphales
Lentaria	Pterulaceae	
Macrotyphula	Clavariadephaceae	
Pterula	Pterulaceae	
Ramaria	Ramariaceae	
Calocera	Dacrymycetaceae	Dacrymycetales
Ditiola		
Eocronartium	Cystobasidiaceae	Platygloeales
Gymnosporangium	Pucciniaceae	Uredinales
Arthrosporella (anamorphic state; see Fig. 88)	Tricholomataceae	Tricholomatales

group to the earth fans *Thelephora*, namely *Polyzellus*, with angular, brown spores (Figs 48–53). The residue is united in the smooth, relatively thin-walled, hyaline spores. The chemistry of these fungi also follows these same groupings.

Once the links between these apparently diverse growth forms were established, some sense could then be made of their relationships resulting in the production of several morphological series, indeed in the case of the gomphoid fungi, even including a resupinate genus *Kavinia* (Fig. 46). In the boletes there has been a loss of ballistosporic dispersal and drift towards the

Figs 43–53 (opposite). Cantharelloid, Ramarioid and Thelephoroid fungi. **Fig. 43a & b.** False chanterelle – *Hygrophoropsis aurantiaca* with spores (43a) and 2 spores of *Cantharellus cibarius* (43b). **Fig. 44.** *Gomphus clavatus* with spore. **Fig. 45.** *Ramaria* spp.: *R. flava* (a), *R. invalii* (b) & *R. zippelii* (c). **Fig. 46.** *Kavinia* sp.; resupinate 'Ramaria'. **Fig. 47.** Selection of spores in *Ramaria*. **Fig. 48.** Selection of spores in *Thelephora*. **Fig. 49.** *Thelephora pseudoterrestris* from S.E. Asia. **Fig. 50.** *T. terrestris* from temperate area. **Fig. 51.** *Tomentella* sp.; resupinate 'Thelephora'. **Fig. 52.** *Thelephora palmata.* **Fig. 53.** *Hydnellum ferrugineum.*

Table 8. *Comparison of the Ganodermatales, an order with only poroid fruit bodies, and the Thelephoraceae and Hymenochaetales, which each show a wide spectrum of hymenophore configurations (indicated by extent of the horizontal bars)*

	Traditional families							
	Agaricaceae	Cantharellaceae	Thelephoraceae	Hydnaceae	Clavariaceae	Polyporaceae		
			Hymenophore configuration					
Modern orders and families	Gilled	Ridged and veined: pileate	Smoth to wrinkled: resupinate to effusso-reflexed	Toothed	Club to coral shaped	Tubes resupinate Tubes		Tubes
						Annual or perennial (*Poria*)	Non-resupinate Annual (*Polyporus*)	Perennial (*Fomes*)
	Agaricoid	Cantharelloid	Thelephoroid	Hydnoid	Clavarioid	Poroid	Poroid	Poroid
Thelephorales								
Thelephoraceae								
Lenzitopsidacea								
Bankeraceae								
Boletopsidaceae								
Verrucosporacea*								
Hymenochaetales								
Clavariachaetac								
Hymenochaetacea								
Colticiaceae								
Phellinaceae								
Ganodermatales								
Ganodermataceae								
Haddowiaceae								

*There is some doubt that *Verrucospora*, the sole represenatvie, belongs to this order. Taxa according to Julich (1981)

Fig. 54. *Leccinum holopus*, a characteristic bolete of wet boreal forest, ectomycorrhizal with a range of *Betula* spp.

production of stigmatospores but this is not the case in all series. Perhaps molecular analysis may be able to help with this problem as has been shown in a limited way by Baura *et al.* (1992). Some examples are offered in Figs 43–53 and Tables 7 & 8; Corner (1966) has illustrated the development of many of the chanterelles.

Dry rot Fungus, a resupinate bolete?

In the discussion earlier the boletes including the ceps of commerce (Fig. 54), have been lumped with the agarics despite the fact that the former have spongy spore-bearing tissues beneath the cap and the agaric an array of gills. In classical literature they were placed with the polypores, as these also possessed pores and tubes, but it soon became obvious that whilst the polypores had a woody habit, and continuously growing margin which often left 'growth zones', the boletes consistently had a putrescent and definite conserved mushroom shape. Some true agarics even have a poroid spore-producing tissue, e.g. *Filoboletus* and *Dictyopanus* (Fig. 10b). From an intermediate position during the early part of the century, the boletes rapidly took their place with the agarics, even being placed with them in a single order, the Agaricales (Singer, 1951, 1962, 1975, 1986). Presently they are separated into an order along with a few lamellate species e.g. *Phylloporus* and some resupinate fungi which possess many characters in common, even their chemistry (Bresinsky & Orendi, 1970).

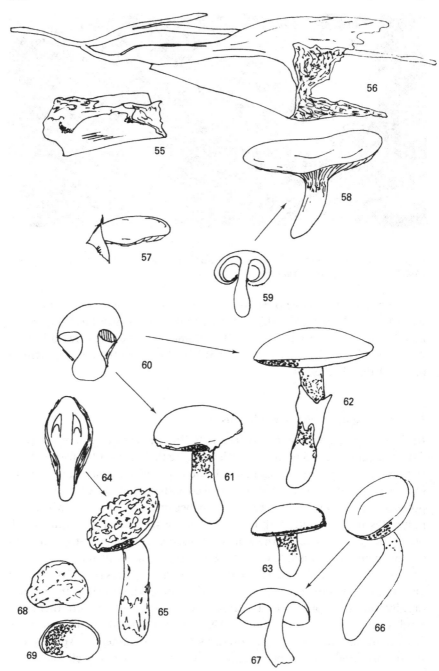

It was, perhaps, the evidence from chemotaxonomy which finally convinced mycologists that there was indeed a link between the boletes and the resupinate dry rot fungus *Serpula lacrimans*, Fig. 56 and the wet rot fungus, *Coniophora puteana* (Fig. 55). For some time anatomical features of the last two fungi had been shown to parallel some of the boletes but as they were resupinate some disbelief was expressed. The link with the boletes as proposed was through the intermediate group which includes the agaricoid *Tapinella* (Fig. 57) and *Paxillus* (Fig. 58). In common with the chemistry there is a parallel in the trophic biology of each of the constituent genera. This latter phenomenon is paralleled in some other series described above.

Coniophora, *Serpula* and *Tapinella* are all gymnocarpic in their development with the spore-producing tissue forming an irregular, lumpy surface in *Coniophora* whereas in *Serpula* both a resupinate and cap-forming (pileate) fruit body may develop depending on the environmental conditions. In common with many of the Aphyllophorales there is great polymorphism in the fruit body.

The change in the shape of boletes attacked by a hypocreaceous fungus has already been alluded to above but it is also interesting to note that such shape change and conversion into masses of spores of *Sepedonium* is only found in a series of fungi, all related by other characters. This has long been known but was placed on a chemical basis by Gill & Watling (1993). Thus, such seemingly diverse forms as *Pisolithus* and *Scleroderma* on the one hand are linked to *Gyroporus* and *Hygrophoropsis* on the other (Watling & Gregory, 1992). Indeed, even the type of the genus *Phyllobolites*, a member of this series, is destroyed by *Sepedonium* (Horak, 1968).

The closely related *Paxillus* includes *P. involutus* which exhibits an extension of the gymnocarpic development. Here, the primordial cap margin grows very close to the stem so protecting the developing gills. Such development is termed pilangiocarpic (Figs 58 & 59), and is also found in several true boletes, e.g. *Suillus variegatus*. In the boletes a whole range of developmental types is found, even examples with enveloping veils. The majority of European boletes are gymnocarpic but those placed in the

Figs 55–67 (opposite). Boletoid fungi and their relatives. **Fig. 55.** *Coniophora puteana*, wet rot fungus. **Fig. 56.** *Serpula lacrimans*, dry rot fungus. **Fig. 57.** *Tapinella panuoides*. **Fig. 58.** *Paxillus involutus*. **Fig. 59.** Young fruit body of *P. involutus* showing enveloping cap (= pileus); pilangiocarpic. **Fig. 60.** Young primordium of *Suillus* spp.; mixangiocarpic. **Fig. 61.** *S. borealis*. **Fig. 62.** *S. luteus*. **Fig. 63.** *S. albivelatus*. **Figs 64 & 65.** Young and mature fruit bodies of *Strobilomyces floccopus*; velangiocarpic. **Figs 66 & 67.** Fruit body and section of *Gastroleccinum scabrosum*; non-gravitropic hymenium. **Figs 68 & 69.** Fruit body and section of *Rhizopogon luteolus*; endocarpic.

Table 9. *Gasteroid boletes: relationships between secotiod and boletoid members of the Boletales*

Boletoid			Secotioid
Gomphidiaceae		*Chroomgomphus*	*Brauniellula*
		Gomphidius	*Gomphogaster*
Paxillaceae		*Paxillus*	*Austrogaster*
		Phylloporus	*Paxillogaster* &
			Gymnopaxillus
Boletaceae		*Boletus*	*Gastroboletus*
	sub-genus	Sect. *Luridi*	e.g. *G. turbinatus*
	Boletus	Sect. *Edules*	e.g. *G. subalpinus*
		Sect. *Calopodes*	e.g. *G. citrinibrunneus*
		Sect. *Mirabiles*	*Truncocolumella rubra*
	sub-genus		*G. xerocomoides*
	Xerocomus		
		Leccinum	*Gastroleccinum*
		Suillus	*Gastrosuillus*
Chamonixiaceae		*Gyroporus?*	*Chamonixia*
Rhizopogonaceae			*Rhizopogon*

genus *Suillus* do show a range of modifications. In some, such as *S. luteus*, the margin of the cap continues to grow after initiation of the spore-bearing tubes to form a membrane tightly adhering to the stem, even swamping the specialised cells on the stem-surface which are so characteristic of other members of the genus. In other species the veil does not hug the stem and is left either as a membrane or roll of tissue at the cap margin, whilst in others the growth of the membrane is so profuse that a huge baggy sock is formed, as in *S. cothurnatus*. But in all cases the basic pattern of the initial development is gymnocarpic; these modifications are termed mixangiocarpic (Figs 60–63).

Secotiaceous members of the boletes follow their boletoid relatives (Table 9). Thus *S. riparius*, although enclosed, has a secondary veil whereas *Gastroboletus turbinatus* and *Gastroleccinum scabrosum* are truly naked (Figs 66 & 67). The ring in the boletes, although superficially similar to the ring in *Agaricus* and *Amanita*, is formed in a very different way and at a very different time in the development of the fruit body.

A further developmental type is found in *Strobilomyces floccopus* where all outer tissues take part in the secondary development of protective membranes. Such development is called metavelangiocarpic (Figs 64 & 65). The young fruit body is enveloped in fungal hyphae which do not keep pace with the expanding fruit body and later disorganise to expose the spore-producing tissue. The development of many tropical species of boletes has been illustrated by Corner (1972). A totally and persistently enveloped relative of the boletes is considered to be the hypogeous genus *Rhizopogon*. This genus reaches its height of speciation in North America where 137 different species are known (Smith & Zeller, 1966; Figs 68 & 69). The metavelangiocarpic development characterises the agaricoid Gomphidiaceae, a family linked to the boletes on chemistry and spore morphology. Also, on the basis of chemical and anatomical features of the ectomycorrhizal sheath, Agerer (1991) has suggested a link between the boletes and the Thelephoraceae (see under chanterelles above).

Tissue structure

Development of the fruit body relies on the way the hyphae composing the fruit body unite in their growth and branching to form the final product. As shown above, many generations of mycologists have slavishly followed a classification based on the final form of the fungus, not taking into account all the available features. This has now changed, but careful observation and critical reappraisal of so-called well-known organisms and the ways in which tissues differentiate is proving informative.

Fayod (1889) demonstrated that there was a fundamental difference between one group of agarics with flesh composed of an intimate mixture of filamentous and inflated hyphal components, now united in the Russulaceae (Figs 70 & 71), and the majority of other agarics with their apparently uniform tissue (Fig. 72). The Russulaceae are characterised by discrete packets or cylinders of rounded cells (spherocytes) separated by filamentous hyphae including those containing secondary metabolites and then called laticiferous hyphae (Fig. 70a). Tissue containing such elements is termed heteromerous and it is speculated that by expansion of the spherocytes expansion of the fruit body can take place. The development of this tissue has been examined by Reijnders (1976) and Watling & Nicoll (1980). Although European species of *Russula* are considered gymnocarpic, the annulate members so characteristic of West and Central Africa (Fig. 71) have not been investigated. Superficially, these are gymnocarpic with secondary veils as in the boletes. Much more work is required as the general pattern of development in the Russulaceae is still unknown. Sararni (1994) is presently studying the veils in European taxa and expanding Singer's observations which necessitate erection of a special section in *Russula* termed Subvelatae to reflect the velar character. The whole of this subject must be reopened. Secotiaceous members are also known in this group.

skeletal

lact.

70

70a

71

72

generative

binding

73

74a

74b

75

76

77

79a

78

79b

80

Corner (1966) introduced the concept of the sarcodimitic (Fig. 74a) and sarcotrimitic (Fig. 74b) structures for certain white-spored agarics in which group this is obviously an area of future study (Pegler, 1996). Redhead (1987) brought fungi with these traits together into a single family, the Xerulaceae. Although Corner (1966) described the development of individual hyphae it is too early to indicate if there is any unifying developmental type between members of the family. Thus *Oudemansiella mucida* is bivelangiocarpic (Fig. 76) whereas *Xerula radicata* is paravelangiocarpic and *Megacollybia platyphylla* and *Mycena rorida* are gymnocarpic (Fig. 75).

The expansion seen in the thin-walled cells found in the Russulaceae may exemplify the mechanism in members of the Amanitaceae which, although generally bulky fungi, still develop rather rapidly. All members are similar in being bivelangiocarpic, and in having the tissue of their gills torn from the background tissue of the fruit body as it matures. In Amanitaceae thin-walled, highly septate hyphae swell rapidly to produce chains of cells. Such structures are called acrophysalids (Fig. 77) and might be hypothesised as being responsible for the rapid development of the fruit body, especially the elongation of the stem, in some cases leaving the ring as a small limb hidden in the volva (Figs 78 & 79a & b).

In *Coprinus cinereus* further research will undoubtedly indicate the role played by the narrow hyphae found in the stem (Hammad, Watling & Moore, 1993; Fig. 80). On the basis of their heterogeneous reaction with all (of many) histological staining reactions applied, Hammad *et al.* (1993) suggested that narrow hyphae fulfil a number of different roles. Communication and supply of nutrients are presumably among these, and since these hyphae form a network around and between the inflated hyphae which form the bulk of the structure of the stem, some sort of mechanical or structural function is also likely (and see Chapter 1). At present it is not known whether these narrow hyphae are general to all agarics, or limited to

Figs 70–80 (opposite). Tissue organization. **Figs 70 & 71.** Russulaceae: section of European *Russula fellea* and African *R. annulata* with heteromerous trama in rosettes and cylinders or columns; (lact.) = laticiferous hyphae (70a). **Fig. 72.** Tricholomataceous agaric with homoiomerous tissue. **Fig. 73.** Trimitic structure of woody polypore including generative, skeletal, and binding hyphae. **Fig. 74a & b.** Sarcodimitic and sarcotrimitic tissue. **Fig. 75.** *Mycena rorida.* **Fig. 76.** *Oudemansiella mucida.* **Fig. 77.** Acrophysalidic tissue in section of *Amanita* gill. **Fig. 78.** Bivelangiocarpic development in *Amanita.* **Fig. 79a & b.** *Amanita fulva* showing in section (79b) position of second veil; homologous to 'partial' veil in *A. muscaria* (in Fig. 6). **Fig. 80.** Tissue of *Coprinus cinereus* showing large and narrow hyphal elements.

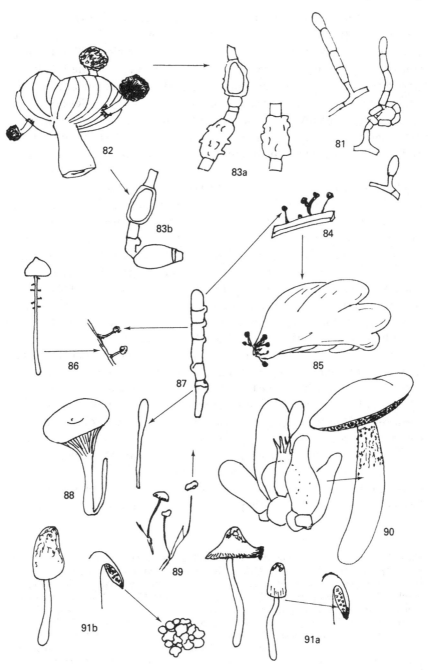

members of the Coprinaceae, to members of the genus *Coprinus*, or even restricted to certain sections of that genus. Certainly, *Coprinus* demonstrates a mixture of developmental types, namely biangiocarpic, monoangiocarpic and gymnocarpic, and DNA studies have reinforced this disparity (Hopple & Vilgalys, 1994).

Anamorphic and teleomorphic structures

Conidial states, some quite complex, are a feature of many ascomycetes and even of some aphyllophoroid basidiomycetes but in the agarics they are very much reduced (Watling, 1979; Fig. 81). However, in the genus *Nyctalis* (*Asterophora*) conidial production dominates and the asexual propagules are borne directly on and within the umbrella-shaped fruit body (Figs 82 & 83a and b). A parallel structure, but one producing packets of sclerotioid cells, is found in *Rhacophyllus lilacinus*, the anamorphic stage of *Coprinus clastophyllus* (Figs 91a and b). Even the stem cortex in some agarics is capable of forming pin-headed structures, termed coremia which bear conidia, for example *Collybia racemosa* (Fig. 86), *Pleurotus cystidiosus* (Fig. 85; anamorph *Antromycopsis*, Fig. 84) and *Arthrosporella ditopa* (Fig. 88)(see Watling & Kendrick, 1977). Such asexual spores may be dispersed by air currents, insects, and probably molluscs and water. In *Mycena citricolor* the whole cap becomes a diaspore (Fig. 89).

Basidia occur on the stem-cortex of those agarics and boletes lacking veils (gymnocarpic) and even in those in which the veil forms after the initiation of spore-production (mixangiocarpic), e.g. *Leccinum* spp. (Fig. 90). Many authors have observed the production of fertile basidia on the pilei of a range of agarics; indeed the outermost layer of the cap termed the pileipellis is composed in several agaric genera of a hymeniform layer with many elements resembling those generally seen on the gill (hymenophore). Kniep (1911) found that basidia were produced on the vegetative mycelium

Figs 81–91 (opposite). Conidial apparatus and parallel phenomena. **Fig. 81.** Thallic arthrospores in Bolbitiaceae. **Fig. 82.** *Nyctalis asterophora.* **Fig. 83a & b.** Conidia of *N. asterophora* (83a) and of *N. parasitica* (83b). **Fig. 84.** *Antromycopsis broussentiae.* **Fig. 85.** *Pleurotus cystidiosus* with *Antromycopsis* stage on stem. **Fig. 86.** *Collybia racemosa* with conidial coremia on stem. **Fig. 87.** Thallic arthrospores in Figs 84 & 86. **Fig. 88.** *Arthrosporella ditopa*, agaricoid and clavarioid forms. **Fig. 89.** *Mycena citricolor* repulsing its cephalodia or 'gemmae' which are modified caps. **Fig. 90.** Basidia in caulohymenium on stem of *Leccinum*. **Fig. 91a & b.** *Coprinus clastophyllus* (91a) with conidial stage *Rhacophyllus lilacinus* (91b) producing pockets of sclerotized cells.

of *Armillaria* in culture, and Bastos & Andebrhan (1987) and Yamanaka & Sagara (1990) have demonstrated the production of basidia in cultures of both *Crinipellis perniciosa*, the causal agent of witches' broom disease of cocoa, and *Lyophyllum tylicolor*, a species which grows on soil highly impregnated with nitrogenous compounds. This ability to produce basidia directly on the mycelium adds a further dimension to the flexibility of certain agarics and expands their capability of dispersal over and above the production of asexual propagules which themselves may be either dikaryotic or monokaryotic, or both. With such material it might be possible to examine this transformation in the laboratory and attempt to come to grips with the biochemical mechanisms involved.

The loss of gills in the cyphelloid fungi refocuses the attention on the gill arrangement in the agarics which has long been used as a taxonomic character. For example, decurrent where the gills run down the stem, adnate where the gills are closely pressed to the stem for their entire width, etc (Fig. 5). and illustrated in most elementary texts and field guides (see Henderson, Orton & Watling, 1969) but in a purely descriptive way without further analysis. It is true that several gill conditions are constant, e.g. the gills of Pluteaceae are always free of the stem, but others are less reliable and indeed may go through change as the fruit body develops, as in many members of the genus *Hygrocybe*.

There is some evidence that the shape of the cap and gill attachment in some species are intimately associated; for example, *Omphalina* with decurrent gills (Fig. 5b) and depressed cap, but there are exceptions. The relationship of the number of gills to cap expansion has been adequately dealt with by Poder (1992) and Caillat, Caillat & Modera (1993). Certainly, some taxa are characterised by consistently close gills, such as *Pluteus* spp., many *Coprinus* spp., and the role of the cystidia in regulating their position is only now being understood (Rosin, Horner & Moore, 1985; Chiu & Moore, 1990a). Some taxa have widely spaced gills and here the cystidia possibly play a greater role by acting as baffles to limit water loss (Watling, 1975).

Conclusion

Only fairly recently has there been an active analysis of the supposed correlations between the final form of the fruit body as seen in the field and spore dispersal patterns of development. It is important to study dispersal mechanisms in agaricoid fungi in a parallel way to the long tradition in phanerogamic studies where pollinators have been systematically recorded

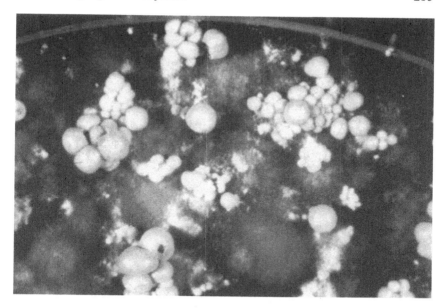

Fig. 92. Gasteroid fruit bodies of a strain of *Psilocybe merdaria* in culture.

in autoecological studies (see, for example, the *Biological Flora* published for some British plants in the *Journal of Ecology*). At last we appreciate that although the classic agaric disperses its sexual spores (basidiospores) by wind currents, relatives rely on insects or mammals for spore dispersal, e.g. hypogeous taxa or even water currents (Desjardin *et al.*, 1995). Perhaps we are only now shaking off the traditional way of viewing macromycetes.

Hibbett *et al.* (1994) have shown in laboratory studies the importance of the correct interpretation of the development of the fruit body for a correct understanding of evolution and classification, something Reijnders' recent observations (Reijnders & Stalpers, 1992) and laboratory work with ribosomal DNA (Hibbett & Vilgalys, 1991) have pointed to. Systematics, by its nature of finding relationships between organisms, leads to a degree of predictability which allows an intelligent choice of material for future studies in our enquiry into fruit body formation. Data so obtained could be of use to the mushroom industry by enabling exploitation of further exotic foods.

In the laboratory several mutants are now available for experimentation, or with some ingenuity can be produced if required, e.g. gastroid forms of *Psilocybe merdaria* (Fig. 92; Watling, 1971). Such isolates represent some of the fruit body forms discussed above, e.g. expansionless mutants of *Cop-*

Fig. 93. *Thelephora terrestris* growing up the stem of a potted *Picea* under glasshouse conditions.

rinus cinereus which fail to expand the cap. Even if these do not represent the actual way by which these morphologies are formed in stabilised populations in nature they nevertheless might give some clues as to the pathways involved which can then be analyzed biochemically and possibly genetically. It should always be remembered, as indicated above, that there exists a great amount of plasticity in many basidiomycetes, e.g. *Volvariella bombycina* (Chiu & Moore, 1990b) and that in many taxa the production of the fruit body can be easily put off course (Watling & Moore, 1994).

Agarics are more plastic in their form than was previously believed and of the author offers a word of caution to the interpretation of field data and its significance in evolutionary studies.

Fig. 94. Unidentified polypore from West Africa showing both pileate and resupinate forms in the same fruit body. Whole fruit body 120 mm, cap 65 mm broad.

Fig. 95. Aberrant form of *Merulius tremellosus* joining up at margins to form turbinate fruit body.

The plasticity exhibited by *Thelephora terrestris* which may engulf plant material or rocky debris as it grows (Fig. 93) and that of an unidentified species of polypore from West Africa which formed a plate-like fruit body on one part of the substrate and a pileate form in another (Fig. 94) shows that variation can also be found in the non-agaricoid fungi. Occasionally, even in these forms the pilei can fuse to form a centrally stipitate fruit body as exhibited by an unusual form of *Merulius tremellosus* Fr. (Fig. 95). This species is generally also effuso-reflexed although other species in the genus are resupinate. The genus *Merulius* in the classical sense is a mixture of

Fig. 96. *Calyptella campanula* on cut tomato stems.

disparate elements even including the familiar causal organism of Dry Rot, *Serpula lacrimans* (Fig. 56), all drawn together by their gross morphology. As seen above, some of these non-agaricoid fungi are more related to the agarics and boletes than some classifications would indicate.

Undoubtedly, careful developmental studies will allow an improved understanding of this huge mix of species known as resupinates. A few have already been referred to in the discussion above but we have a long way to go in placing many of the others. Although mating tests in pure culture are routine, no concerted effort has been made to fruit many of them in pure culture. This will be a promising line of enquiry in the future, particularly as some are considered advanced forms whilst others are considered primitive.

However, one of the beauties of the cyphelloid and many tricholomataceous fungi other than their fascinating morphology is that they are saprotrophs and have great potential as experimental material (Fig. 96). Bolbitiaceous, coprinaceous and strophariaceous agarics on the other hand offer a possible easy entry into an experimental analysis of at least some aspects of the gasteroid fruit body.

Acknowledgements

It is with great pleasure I record the assistance given by my colleague Evelyn Turnbull, encouragement from and collaboration with David Moore, and discussions with A. F. M. Reijnders and the late M. A. Donk.

References

Agerer, R. (1983). Typusstudien an Cyphelloiden Pilzen IV *Lachnella* Fr. s.l. *Mitteilungen aus der Botanische Staatssammlung München* **19**, 163–334.

Agerer, R. (1991). Studies on ectomycorrhizae XXXIV. Mycorrhizae of *Gomphidius glutinosus* and of *G. roseus* with some remarks on Gomphidiaceae. *Nova Hedwigia* **53**, 127–170.

Bastos, C. N. & Andebrhan, T. (1987). *In vitro* production of basidiospores of *Crinipellis perniciosa* the causative agent of witches' broom disease of cocoa. *Transactions of the British Mycological Society* **88**, 406-409.

Baura, G., Szaro, T. M. & Bruns, T. D. (1992). *Gastrosuillus laricinus* is a recent derivative of *Suillus grevillei*. *Mycologia* **84**, 592–597.

Bougher, N. L., Tommerup, I. C. & Malajcuk, N. (1993). Broad variation in developmental and native basidiome morphology of the ectomycorrhizal fungus *Hydnangium sublamellatum* sp. nov. bridges morphologically based generic concepts of *Hydnangium*, *Podohydnangium* and *Laccaria*. *Mycological Research* **97**, 613–619.

Bresinsky, A. & Orendi, P. (1970). Chromatographische Analyse von Farbmerkmalen der Boletales und anderer Macromyzeten auf Dnnschichten. *Zeitschrift für Pilzkunde* **36**, 135–169.

Caillat, B., Caillat, J. & Modera, J. (1993). L'Induce lamellaire ou essai sur la quantification des lamelles. *Bulletin trimestriel de la Fédération Mycologique Dauphiné-Savoie* **131**, 20–24.

Cannon, P. F., Hawksworth, D. L. & Sherwood-Pike, M. A. (1985). *The British Ascomycotina. An Annotated Checklist*. CAB International: Wallingford, U.K.

Chiu, S. W. & Moore, D. (1990a). A mechanism for gill pattern formation in *Coprinus cinereus*. *Mycological Research* **94**, 320–326.

Chiu, S. W. & Moore, D. (1990b). Development of the basidiome of *Volvariella bombycina*. *Mycological Research* **94**, 327–337.

Clark, W. S., Richardson, M. J. & Watling, R. (1983). Calyptella root rot – a new fungal disease of tomatoes. *Plant Pathology* **32**, 95–99.

Corner, E. J. H. (1964). *The Life of Plants*. Weidenfeld & Nicholson: London.

Corner, E. J. H. (1966). A monograph of the cantharelloid fungi. *Annals of Botany*, Memoirs No. **2**, 1–255.

Corner, E. J. H. (1972). *Boletus in Malaysia*. Singapore Government Printer: Singapore.

Corner, E. J .H. (1993). 'I am a part of all that I have met' (Tennyson's Ulysses). In *Aspects of Tropical Mycology* (ed. S. Isaac, J. C. Frankland, R. Watling & A. J. S. Whalley), pp. 1–13. Cambridge University Press: Cambridge, U.K.

Dennis, R. W. G., Orton, P. D. & Hora, F. B. (1960). New Check list of British Agarics and Boleti. *Transactions of the British Mycological Society* **40** (supplement), 1–225.

Desjardin, D. E., Martinez-Peck, L. & Rajchenberg, M. (1995). An unusual psychrophilic aquatic agaric from Argentina. *Mycologia* **87**, 547–550.

Dissing, H. (1966). The genus *Helvella* in Europe. *Dansk Botanisk Arkiv* **25** (1), 1–172.

Donk, M. A. (1966). A reassessment of the Cyphellaceae. *Acta Botanica Neerlandica* **15**, 95–101.

Fayod, V. (1889). Histoire Naturelle Des Agaricinés. *Annales des Sciences Naturelles Septième Série, Botanique* **9**, 181–411.

Gill, M. & Watling, R. (1986). The relationships of *Pisolithus* (Sclerodermataceae) to other fleshy fungi with particular reference to the occurrence and taxonomic significance of hydroxylated pulvinic acids.

Plant Systematics and Evolution **154**, 225–236.

Hammad, F., Watling, R. & Moore, D. (1993). Cell population dynamics in *Coprinus cinereus*: narrow and inflated hyphae in the fruit body stem. *Mycological Research* **97**, 275–282.

Henderson, D. M., Orton, P. D. & Watling, R. (1969). *British Fungus Flora, Agarics & Boleti.* HMSO: Edinburgh.

Hibbett, D. S., Grimaldi, D. & Donoghue, D. J. (1995). Cretaceous mushroom in amber. *Nature* **377**, 487.

Hibbett, D. S. & Vilgalys, R. (1991). Evolutionary relationships of *Lentinus* to the Polyporaceae: evidence from restriction analysis of enzymatically amplified ribosomal DNA. *Mycologia* **83**, 425–439.

Hibbett, D. S., Shigeyuke, M. & Tsumeda, A. (1994). Sporocarp ontogeny in *Panus*. Evolution classification. *American Journal of Botany* **80**, 1336–348.

Hopple, J. S. Jr. & Vilgalys, R. (1994). Phylogenetic relationship among coprinoid taxa and allies based on data from restriction site mapping of nuclear DNA. *Mycologia* **86**, 96–107.

Horak, E. (1968). Synopsis generum Agaricalium. *Beitrage zur Kryptogamenflora der Schweiz* **13**, 1–741.

Horak, E. & Desjardin, D. E. (1994). Reduced marasmioid and mycenoid agarics from Australia. *Australian Systematic Botany* **7**, 153–170.

Jonsson, L. & Nyland, J. E. (1979). *Favolaschia dybowskyana* (Singer) Singer, a new orchid mycorrhizal fungus from tropical Africa. *New Phytologist* **83**, 121–28.

Julich, W. (1981). Higher taxa of Basidiomycetes. *Bibliotheca Mycologia* **85**, 1–485.

Kniep, H. (1911). ber das Auftreten von Basidien im einkernigen Myzel von *Armillaria mellea* Fl. Dan. *Zeitschrift für Botanik* **3**, 529–533.

Knudsen, H. (1995). Taxonomy of the basidiomycetes in Nordic Macromycetes. *Symbolae Botanicae Upsaliensis* **30** (3), 169–208.

Moore, D., Hock, B., Greening, J. P., Kern, V. D., Novak Frazer, L. & Monzer, J. (1996). Centenary review. Gravimorphogenesis in agarics. *Mycological Research* **100**, 257–273.

Patouillard, N. (1900). *Essai Taxonomique sur les Familles et les Genres des Hyménomycètes.* Lucien Declume: Lons-Le-Saunier.

Pegler, D. N. (1996). Centenary review. Hyphal analysis of basidiomata. *Mycological Research* **100**, 129-142.

Poder, R. (1992). Aspects of gill development and proportions in basidiocarps. *Mycologia Helvetica* **5**, 39–46.

Redhead, S. A. (1987). The Xerulaceae (Basidiomycetes) a family with sarcodimitic tissues. *Canadian Journal of Botany* **65**, 1551–1562.

Reijnders, A. F. M. (1948). *Etudes sur le développement et l'organisation histologique des carpophores dans les Agaricales. Extract du Recueil des Travaux botaniques néerlandais*, **Vol 4**, 213–396.

Reijnders, A. F. M. (1963). *Les Problèmes du Dèveloppement des Carpophores des Agaricales et de Quelques Groupes Voisins.* W. Junk: The Hague.

Reijnders, A. F. M. (1976). Recherches sur le développement et l'histogénèse dans les Asterosporales. *Persoonia* **9**, 65–83.

Reijnders, A. F. M. & Stalpers, J. A. (1992). The development of the hymenophoral trama in the Aphyllophorales and the Agaricales. *Studies in Mycology* **34**, 1–109.

Rosin, I. V., Horner, J. & Moore, D. (1985). Differentiation and pattern formation in the fruit body cap of *Coprinus cinereus*. In *Developmental Biology of Higher Fungi* (ed. D. Moore, L. A. Casselton, D. A. Wood, & J. C. Frankland), pp. 333–351. Cambridge University Press: Cambridge, U.K.

Sararni, M. (1994). *Russula* nuovo o interessanti dell'Italia centrale e mediterranea 23° contrib. problemi di specificazione e di sistemazione delle russule europe e velangiocarpe. *Micologia Italiana* **23**, 27–36.

Singer, R. (1951). Agaricales in Modern taxonomy. *Lilloa* **22** (1949) 1–830.

Singer, R. (1958). The meaning of the affinity of the Secotiaceae with the Agaricales. *Sydowia Annales Mycologici* Series II, **12**, No. 1–6, 1–43.

Singer, R. (1962). *The Agaricales in Modern Taxonomy*, 2nd edn. J. Cramer: Weinheim.

Singer, R. (1975). *The Agaricales in Modern Taxonomy*, 3rd edn. J. Cramer: Vaduz.

Singer, R. (1986). *The Agaricales in Modern Taxonomy*, 4th edn. Koeltz Scientific Books: Koenigstein.

Singer, R. & Smith, A. H. (1960). Studies on secotiaceous fungi. IX The Astrogastraceous series. *Memoirs of the Torrey Botanical Club* **21**, 1–112.

Smith, A. H. & Zeller, S. M. (1966). A preliminary account of the North American species of *Rhizopogon*. *Memoirs of the New York Botanic Garden* **14**, 1–177.

Thiers, H. D. (1984). The secotioid syndrome. *Mycologia* **76**, 1–8.

Watling, R. (1971). Polymorphism in *Psilocybe merdaria*. *New Phytologist* **70**, 307–326.

Watling, R. (1975). Studies in fruit body development in the Bolbitiaceae and the implications of such work in studies on higher fungi. *Beihefte Nova Hedwigia* **51**, 319–346.

Watling, R. (1978). From Infancy to adolescence, advances in the study of higher fungi. *Transactions of the Botanical Society of Edinburgh* (Supplement) **42**, 61–73.

Watling, R. (1979). The morphology, variation & ecological significance of anamorphs in the Agaricales. In *The Whole Fungus, Kananaskis* II. pp. 453–472. National Museum of Canada & Kananaskis Foundation: Ottawa, Canada.

Watling, R. (1985a). Developmental characters of agarics. In *Developmental Biology of Higher Fungi* (ed. D. Moore, L. A. Casselton, D. A. Wood & J. C. Frankland), pp. 281–310. Cambridge University Press: Cambridge, U.K.

Watling, R. (1985b). Hymenial surfaces in developing agaric primordia. *Botanical Journal of the Linnean Society* **91**, 273–293.

Watling, R. & Gregory, N. M. (1992). Observations on the boletes of the Cooloola sand-mass Queensland and notes on the distribution in Australia. Part 3, Lamellate taxa. *Edinburgh Journal of Botany* **48**, 353–393.

Watling, R. & Kendrick, B. (1977). Dimorphism in *Collybia racemosa*. *Michigan Botanist* **16**, 65–72.

Watling, R. & Moore, D. (1994). Moulding moulds into mushrooms: shape and form in the higher fungi. In *Shape and Form in Plants and Fungi* (ed. D. S. Ingram & A. Hudson), pp. 271–290. Academic Press: London.

Watling, R. & Nicoll, H. (1980). Sphaerocysts in *Lactarius rufus*. *Transactions of the British Mycological Society* **75**, 331–333.

Yamanaka, T. & Sagara, N. (1990). Development of basidia and basidiospores from slide-cultured mycelia in *Lyophyllum tylicolor* (Agaricales). *Mycological Research* **94**, 847–850.

Index